Pickling and Fermentation for the Prepared

Creating a Resilient Food Supply

Grace Tanaka

© Copyright 2024 - All rights reserved.

The content contained within this book may not be reproduced, duplicated or transmitted without direct written permission from the author or the publisher.

Under no circumstances will any blame or legal responsibility be held against the publisher, or author, for any damages, reparation, or monetary loss due to the information contained within this book, either directly or indirectly.

Legal Notice:

This book is copyright protected. It is only for personal use. You cannot amend, distribute, sell, use, quote or paraphrase any part, or the content within this book, without the consent of the author or publisher.

Disclaimer Notice:

Please note the information contained within this document is for educational and entertainment purposes only. All effort has been executed to present accurate, up to date, reliable, complete information. No warranties of any kind are declared or implied. Readers acknowledge that the author is not engaging in the rendering of legal, financial, medical or professional advice. The content within this book has been derived from various sources. Please consult a licensed professional before attempting any techniques outlined in this book.

By reading this document, the reader agrees that under no circumstances is the author responsible for any losses, direct or indirect, that are incurred as a result of the use of information contained within this document, including, but not limited to, errors, omissions, or inaccuracies.

Table of Contents

INTRODUCTION

In today's quickly transforming world, the significance of self-reliance and preparedness has never been more apparent. The need for resilient food supplies is paramount as global uncertainties continue to unfold, from economic fluctuations to climate instability and unforeseen emergencies. In this context, the age-old practices of pickling and fermentation emerge as invaluable skills for individuals and communities seeking to secure their food sources and adapt to changing circumstances.

Welcome to "Pickling and Fermentation for the Prepared: Creating a Resilient Food Supply," a comprehensive guide designed to equip you with the knowledge as well as skills necessary to harness the power of pickling and fermentation in your own kitchen. In the pages that follow, we will undertake on a journey through the rich history, practical techniques, and transformative potential of these time-honored food preservation methods.

At its core, this book is about more than just preserving fruits, vegetables, and other perishables. It's about embracing a mindset of resilience and resourcefulness—a mindset that empowers individuals to take control of their food supply, reduce waste, and cultivate a deeper connection to the foods they consume. Whether you're a seasoned home cook looking to expand your culinary repertoire or a novice eager to explore the world of food preservation, there's something here for everyone.

Before we delve into the intricacies of pickling and fermentation, let's take a moment to understand why these practices are more relevant now than ever before. In an era marked by food insecurity, environmental degradation, and supply chain disruptions, producing and preserving one's own food offers a sense of security as well as stability that is increasingly rare in today's fast-paced world. By learning the way to pickle and ferment foods, you're not just acquiring a set of skills but investing in your own well-being and that of your community.

Pickling and fermentation have been used for centuries by cultures around the world to extend the shelf life of perishable foods, enhance their flavor and nutritional value, and create unique culinary delights. From the tangy crunch of pickled cucumbers to the probiotic-rich goodness of homemade sauerkraut, the possibilities are as endless as your imagination.

In the chapters ahead, we'll explore the fundamentals of pickling and fermentation, from understanding the science behind these processes to mastering the techniques and recipes that will alter your kitchen into a hub of creativity and sustainability. Whether you're interested in preserving the bounty of your garden, reducing food waste, or simply exploring new culinary horizons, you'll find practical tips, step-by-step instructions, and mouthwatering recipes to guide you every step of the way.

But this book is about more than just pickles and kraut. It's about resilience in the face of uncertainty, reclaiming control over our food systems, and building stronger, more sustainable communities. It's about recognizing the interconnectedness of food, health, and the environment and taking meaningful action to shape a future that is nourishing, equitable, and resilient for all.

So, join me on this journey as we unlock the secrets of pickling and fermentation and discover the transformative power of resilient food preparation. Together, we'll learn how to create a more sustainable, secure, and satisfying future—one jar of pickles at a time.

CHAPTER I

Understanding Pickling and Fermentation

Overview of Pickling and Fermentation

Two methods of food preservation that have been used for thousands of years all over the world are pickling and fermentation. Both of these methods are considered to be old. Not only do both procedures preserve food, but they also improve its flavor and the nutritional content of the ingredients. This section offers a comprehensive overview of pickling and fermentation, focusing on their respective histories, techniques, and benefits, as well as the various cultural activities that are associated with both methods.

There is evidence that suggests that ancient Mesopotamians may have been pickling as early as 2400 B.C., which is evidence that the process of pickling dates back to ancient times. Pickling is known as a method of preserving food that includes submerging it in an acidic solution, typically vinegar, or through lacto-fermentation, which involves the addition of salt to suck away water and create an environment in which only bacteria that produce lactic acid can flourish. This is because the acidic atmosphere inhibits the growth of germs that cause food to go bad, which in turn helps to preserve the food. Pickling is a typical method of preserving foods such as vegetables, fruits, eggs, and meats. There are a variety of pickling recipes and traditions that are specific to each country. Some examples are the spiciness of kimchi from Korea, the tanginess of sauerkraut from Germany, and the sweetness of pickles from the United States.

On the other hand, fermentation is a process that decomposes food components like carbohydrates into alcohol, gasses, or organic acids. This is accomplished by the action of microorganisms, such as bacteria, yeasts, and molds, which are responsible for the fermentation process. This transition, which is biochemical in nature, not only helps to preserve the food but also has the potential to enhance its nutritional profile and digestibility. The process of fermentation has played a significant role in human nutrition, resulting in the production of a diverse range of fermented foods and beverages. These include bread, cheese, yogurt, wine, beer, and soy sauce, amongst others. There are a number of factors that can have an effect on the fermentation process. These include temperature, salinity, and the presence of particular microbes. These parameters vary greatly depending on the food product and the cultural methods that are being utilized.

In addition to preservation of food, pickling and fermentation both offer substantial benefits to the general public. Through the introduction of a wide variety of flavors and textures, they add to the diversity of the human diet, so enriching the culinary traditions of others. In addition, fermented and pickled foods are well-known for their wellness benefits, which include enhanced digestion and improved gut health. The presence of probiotics, which are helpful bacteria that can be found in an array of fermented foods, has the potential to improve the microbiome of the gut, so bolstering the immune system and possibly lessening the risk of a number of diseases.

In addition, these preservation techniques are both sustainable and kind to the environment. By allowing the preservation of seasonal produce for extended periods of time, they help to reduce the amount of food that is wasted. Pickling and fermentation are processes that are accessible and practical for communities all over the world. While current preservation techniques such as freezing and canning demand a significant amount of energy, pickling and fermentation take extremely little energy.

Both pickling and fermentation are considered to be of great cultural significance in a number of different countries. There are several instances in which these methods are strongly ingrained in the celebrations and traditions of a particular culture. For instance, the preparation and consumption of sauerkraut is a wintertime custom that has been practiced for a very long time throughout Eastern Europe. In Asia, and notably in Korea, kimchi is not merely a side dish; rather, it is a national symbol that is celebrated and incorporated into regular meals. For the most part, the preparation of these delicacies entails activities and rituals that are performed collectively, which serve to enhance social bonds and cultural identity.

There is still a place for pickling and fermentation in contemporary cooking, despite the fact that they have ancient roots. Both professional chefs and amateur cooks explore and develop within these traditions, resulting in the creation of novel flavors and combinations that represent contemporary tastes while still paying homage to the historical foundations of the methods under consideration. Fermented foods are experiencing a rebirth in popularity all over the world, which is part of a larger interest in traditional and artisanal food practices. This interest highlights a need for foods that are not only delectable and nutritious, but also sustainable and anchored in cultural heritage.

In conclusion, pickling and fermentation are two of the most fundamental examples of food preservation techniques that have been extremely important to the development of human civilization. Not only do these methods guarantee the availability of food, but they also enhance our meals by means of a wide variety of flavors and nutrients. It is important to note that the cultural relevance of fermented and pickled foods highlights the profound ties that exist between food, tradition, and community. As we continue to investigate and develop new approaches within these time-honored techniques, we are paying homage to the knowledge of our predecessors while also making a contribution to a future that is both sustainable and healthy. Through the continued appreciation and practice of pickling and fermentation, we are able to keep a critical relationship to our collective culinary legacy, which is one that continues to develop and inspire generations to come.

Importance of Resilient Food Supply

When it comes to guaranteeing food security and stability in both local and global contexts, having a food supply that is resilient is absolutely necessary. When it comes to providing the nutritional needs of populations, supporting economic stability, and mitigating the effects of environmental, social, and economic shocks, resilience in food supply chains is essential. One way to define resilience is as the ability of a food system to withstand shocks and recover after that disturbance has occurred. Over the course of the past few years, the significance of incorporating resilience into food systems has been increasingly apparent. This is a direct result of a wide variety of issues, which include climate change, natural disasters, geopolitical conflicts, and pandemics.

A robust food supply is of the utmost importance for a number of reasons, one of the most important of which being the direct connection it has with food security. When all people, at all times, have physical, social, as well as economic access to sufficient, secure, and nutritious food that satisfies their dietary needs and food choices for an active as well as healthy life, then food security has been established. Even in the face of interruptions such as catastrophic weather catastrophes, crop failures, or transportation breakdowns, a resilient food supply chain assures that food will continue to be plentiful, inexpensive, and accessible to consumers. Food systems that are resilient can protect themselves from localized shocks and minimize food shortages or price spikes that disproportionately affect disadvantaged groups. This is accomplished by diversifying the sources of production, delivery, and storage utilized by the food system.

Furthermore, a robust food supply is essential for promoting economic stability and livelihoods, particularly in rural communities where agriculture serves as the principal source of income. This is especially true within the context of rural communities. Farmers and producers are better able to preserve their livelihoods and contribute to the broader economic growth of their regions when food supply networks are resilient. This is because they are better equipped to deal with variations in the market and demands from the outside world. Additionally, resilient food systems have the potential to sustain local economies by encouraging entrepreneurial endeavors, the development of new jobs, and activities that add value, such as the operation of food processing and distribution facilities. Communities have the ability to improve their resistance to the effects of external shocks by investing in infrastructure, technology, and capacity building. At the same time, this can also increase the economic prospects available within the food industry.

Furthermore, the significance of a resilient food supply goes beyond the simple provision of nutrition and encompasses broader social and environmental implications as well. Resilient food systems are built on the foundation of sustainable agriculture practices, which include the preservation of biodiversity, the water resources management, and also the conservation of soil. It is possible for sustainable agriculture to improve the long-term productivity and viability of food production while simultaneously limiting negative environmental impacts such as soil degradation, water pollution, and habitat loss. This is accomplished by increasing ecosystem health and resilience throughout the agricultural process. In addition, resilient food systems place a high priority on social fairness and inclusivity, making certain that every member of society has access to nourishing food and chances to participate in the economy. Resilient food systems have the potential to contribute to social cohesion and eliminate inequities in food availability and outcomes. This can be accomplished by empowering underrepresented groups, promoting gender equality, and encouraging community engagement.

There has never been a time when the need for resilient food supply chains has been more critical than it is now, given the emergence of new issues such as climate change, population increase, and urbanization. The variability of the climate and the occurrence of extreme weather events offer major hazards to agricultural productivity, which can disrupt supply chains and exacerbate food insecurity in countries that are already particularly vulnerable. Additionally, existing food systems are being put under strain as a result of increased urbanization and changes in dietary habits. Additionally, creative ways to production, distribution, and consumption are required in order to meet the demands of these systems. Furthermore, the COVID-19 pandemic has brought to light the precarious nature of global supply systems, hence emphasizing the significance of local resilience and self-sufficiency in the production and distribution of food.

In a world that is becoming more linked and uncertain, it is necessary to have a food supply that is resilient in order to promote food security, economic stability, and sustainability. Communities have the power to lessen the effects of shocks and pressures while simultaneously guaranteeing that everyone has equal access to nutritious food if they make investments in resilient food systems that value variety, sustainability, and inclusivity. It is possible for us to construct a more resilient food future that will provide nourishment to people, sustain livelihoods, and safeguard the planet for future generations if we work together, innovate, and take action on policy at the local, national, and international levels.

CHAPTER II

Understanding Pickling

What is Pickling?

The process of pickling is both an art and a science in the world of cuisine. It is a method of preserving food that has been around for thousands of years and has been an essential component of human civilization's ability to store food beyond the growing season. Not only does this method, which has deep roots in the annals of history, culture, and gastronomy, include preserving food in an acidic medium or by fermentation, it not only extends the shelf life of the food but also improves its flavor, nutritional content, and digestibility. As we go deeper into the heart of pickling, it becomes clear that this method of preservation is more than just a way to prevent food from going bad; it is a rich tradition that has developed over the course of millennia, representing the inventiveness and cultural history of nations all over the world.

Pickling, at its core, is the technique of extending the utility of food by modifying its original state by the application of acidity or fermentation. The most frequent pickling agents are vinegar, which is a strong acid that creates a setting that is hostile to the growth of microorganisms, and salt, which encourages fermentation by pulling water out of the food and forming brine. Both of these methods are used to preserve food. Lacto-fermentation is a form of fermentation that results in the production of lactic acid, which functions as a natural preservative. This brine makes it possible for beneficial bacteria to proliferate, particularly in the process of lacto-fermentation. Because of these procedures, the growth of hazardous bacteria is considerably inhibited, which guarantees that the food will continue to be safe to consume for a length of time that is significantly longer.

There is evidence that ancient civilizations in Mesopotamia, India, China, and Egypt preserved their vegetables and fruits by pickling them. This indicates that the history of pickling is as old as agriculture itself. Not only did the method make it possible to store an abundance of vegetables, but it also made it possible to make vitamins and minerals accessible during times of scarcity or when the seasons were changing. The practice of pickling developed around the world over the course of several centuries, with each culture creating its own distinctive techniques and flavors. For example, the spicy kimchi of Korea, the sour pickles of Eastern Europe, and the tangy achar of India are all examples of pickling. The versatility and universality of pickling are brought to light by these various approaches, which also show the relevance of pickling in the culinary traditions of many countries.

There are other advantages to pickling that go beyond preservation. Certain processes, in particular fermentative pickling, have been identified for their positive effects on health. This is mostly due to the presence of probiotics, which are helpful bacteria that support the health of the gut. Probiotics, which are found in abundance in foods like sauerkraut and kimchi, have been associated with a number of health benefits, which includes enhanced immune function, improved digestion, and a decreased chance of developing certain chronic diseases. It is also possible for the pickling process to boost the bioavailability of particular elements, which not only makes these items safe to consume but also makes them advantageous from a nutritional standpoint.

Aside from the positive effects on one's health, pickling also has positive effects on the environment. Through the process of preserving food, pickling helps to cut down on food waste and makes it possible to make effective use of produce that would otherwise go bad. Traditional methods, such as pickling, provide an alternative to current methods of food preservation, which frequently require refrigeration or freezing, that is low in energy consumption and friendly to the environment. This is especially important in a world that is struggling with issues of food security and sustainability. When considered in the context of contemporary attempts to promote sustainable food practices and reduce the carbon footprint associated with food production and storage, this particular feature of pickling is particularly pertinent. In

recent years, there has been a resurgence in interest in pickling, which is reflective of a larger trend toward artisanal and traditional methods of food preparation.

Pickling has gained appeal not only for its practical benefits but also as a form of culinary expression. This is partially due to the fact that people are becoming more conscious of what they eat and are looking for ways to reconnect with the sources of their food. Pickling is a process that is used by both professional chefs and home cooks equally, during which they experiment with novel ingredients, techniques, and flavor combinations. In this revival of pickling, it is not enough to simply revisit old recipes; rather, it is also about pushing the boundaries of this time-honored process in order to produce novel and interesting culinary experiences.

In spite of the fact that it has been around for a very long time and has been subjected to many different variants, the fundamental principle of pickling is still the transformational power of fermentation and acidity. It doesn't matter if you're looking for the tangy crunch of a dill pickle, the spicy kick of kimchi, or the nuanced tastes of a pickled beet; the charm of pickling lies in its capacity to transform ordinary items into something exceptional. The capacity of our ancestors to devise ways of food preservation that have stood the test of time and continue to contribute to the continued enrichment of our diets and cultures is a credit to their inventiveness.

In conclusion, pickling is a diverse culinary activity that covers a rich tapestry of history, culture, science, and art. Pickling is a method that has been around for a long time. Providing a sustainable method to extend the shelf life of perishable foods while simultaneously boosting their flavor and nutritional content, it is a monument to the inventiveness of humans in the field of food preservation. In spite of the fact that we are looking to the future, the enduring history of pickling continues to be an active and developing component of our culinary heritage. It serves as a bridge between the past and the present and paves the way for future developments. Irrespective of whether it is embraced for its health advantages, environmental sustainability, or simply the delight of creating and savoring delectable foods, pickling continues to be an essential and treasured component of our global food landscape.

History of Pickling

The history of pickling is a complex tapestry that has been woven over millennia of human culture. For example, there is evidence that foods that have been pickled date back thousands of years. The demand to extend the shelf life of perishable items and provide a steady food supply led to the development of pickling as a method of preservation. This was especially true in locations where seasonal harvests were plentiful but only lasted for a limited period of time. Archaeological evidence suggests that early civilizations such as the Mesopotamians, Egyptians, and ancient Greeks practiced some type of food preservation by salting, fermenting, or brining. Although it is impossible to pinpoint the exact beginnings of pickling, it is possible that these civilizations were interested in preserving food in some way.

Depending on the materials that were readily available in the area and the conditions of the environment, ancient civilizations evolved a variety of pickling techniques. Probably as a method of preserving extra vegetables like cucumbers, onions, and cabbage, pickling was first introduced in Mesopotamia, which is considered to be one of the cradles of civilization throughout history. It is well known that the ancient Egyptians preserved a broad variety of foods by pickling them in vinegar and salt brine. The Greeks and Romans also utilized vinegar-based pickling techniques to preserve a variety of foods, including olives, grapes, and fish.

The art of pickling was passed down from generation to generation through trade routes and cultural contacts, and it had a significant impact on the culinary traditions of Europe, Asia, and other regions. The practice of pickling became an essential component of food preservation in medieval Europe, particularly during the lengthy winter months when fresh produce was in short supply. Monasteries had a vital part in the creation and dissemination of pickling techniques. Monks experimented with a variety of spices, herbs, and flavorings in order to create various pickled delights that were unique to themselves.

On the Age of Exploration, which occurred between the 15th and 16th centuries, European sailors relied on pickled foods to sustain them on their lengthy sea journeys. Some examples of these meals include sauerkraut, pickled herring, and salted meats for example. In addition to preserving food, pickling was also effective in preventing scurvy, which is a vitamin C shortage that is frequently experienced by sailors who are on extended excursions. It was the broad adoption of pickling techniques by European explorers that contributed to the spread of pickled foods around the globe and the exchange of culinary traditions between different continents.

For millennia, pickling has been an essential component of the culinary traditions of Asia. Over the course of this time, each region of Asia has developed its own distinctive techniques and flavors of pickled ingredients. Traditional Japanese cuisine includes a broad variety of vegetables, fruits, and shellfish that are pickled in rice vinegar, soy sauce, or salt. Pickling, also known as "tsukemono," is a fundamental component of traditional Japanese meals. The term "kimchi" refers to a kind of fermented pickles that are used in Korean cuisine. These pickles are created from vegetables like cabbage, radishes, and cucumbers, and they are fermented with chili peppers, garlic, and other flavors.

In more recent times, there has been a resurgence of interest and appeal in pickling. This is primarily due to the increased enthusiasm for artisanal and traditional methods of food preparation. Experimenting with pickling a wide variety of vegetables, from traditional cucumbers and onions to unusual fruits and spices, is becoming increasingly popular among both home cooks and professional chefs. Pickled foods are typically low in calories and high in vitamins, minerals, and probiotics, which has contributed to the resurgence of interest in pickling. This desire for healthier and more sustainable food options has also been a driving force behind the rebirth of pickling.

Pickling is a process that is still evolving and adapting to modern tastes and preferences. Innovative variations on old recipes and an emphasis on locally sourced and seasonal ingredients are commonplace in the pickling industry today. The ancient skill of pickling is still as active and relevant as it was in the past in today's culinary scene, as seen by the fact that trendy pickle bars offer a variety of artisanal pickles and gourmet restaurants incorporate pickled ingredients into their menus. The history of pickling serves as a reminder of the continuing attraction of preserving food through time-honored processes and the cultural relevance of pickled foods in communities all over the world. This is something that we should keep in mind as we look to the future.

Benefits of Pickling

In addition to preserving food, pickling provides a wide range of other advantages. Pickling is a flexible culinary method that not only lengthens the shelf life of perishable ingredients but also enhances its flavor, texture, and nutritional value. Pickling is a technique that may be used in a variety of different ways. delicacies that have been pickled have earned a beloved position in culinary traditions all across the world. These delicacies, which range from crunchy cucumbers to acidic kimchi, are celebrated for their distinct flavor and their adaptability in the preparation of dishes. The capacity of pickling to transform ordinary products into unique culinary treats that can be enjoyed throughout the year is one of the most significant advantages of this method of preserved food preparation.

A big advantage of pickling is that it helps preserve food, which is one of the most important benefits. Through the process of pickling, which involves submerging components in a solution of vinegar or brine, an acidic environment is produced. This environment prevents the growth of dangerous bacteria, molds, and yeast. The perishable goods' shelf life can be effectively extended with the use of this preservation technology, which enables the meals to be stored for weeks, months, or even years without going bad. Because of this, pickled foods offer a practical and cost-effective method of consuming seasonal components beyond the harvest season. This helps to minimize the amount of food that is wasted and ensures that there is a consistent supply of food.

In addition to preserving food, pickling also improves the flavor and texture of the food, which results in a different flavor profile that is characterized by a combination of sour and salty flavors. Pickled foods are a flexible component that may be used in a variety of culinary creations because of the combination of acidity, saltiness, sweetness, and spice that they contain. This combination gives dishes a sense of depth and complexity. Pickled foods provide a blast of flavor that enlivens the palate and satisfies the senses for a variety of reasons, including the fact that they can be eaten as a crunchy snack, a zesty condiment, or a savory addition to a meal.

In addition, pickling can bring about an increase in the nutritional value of foods by preserving the natural vitamins, minerals, and also antioxidants that are present in them. Although the process of pickling may result in the loss of some water-soluble vitamins, which includes vitamin C, foods that have been pickled maintain a significant number of the nutrients that they originally contained, namely fiber, potassium, and calcium. Additionally, the fermentation process that is engaged in certain forms of pickling can improve the bioavailability of nutrients and boost gut health by introducing helpful probiotic bacteria into the digestive tract. This is accomplished thanks to the fermentation process. Therefore, when consumed in moderation, foods that have been pickled has the potential to be a tasty and nutritious supplement to a diet that is balanced.

A further advantage of pickling is that it may be adapted to a variety of culinary traditions and nutritional requirements, making it a versatile and adaptable method. There is a massive amount of variation in pickling processes throughout different cultures and places, with each tradition offering its own distinctive flavors, ingredients, and methods. Pickled foods, such as Korean kimchi, German sauerkraut, and Japanese tsukemono, are a reflection of the diversity of cuisines around the world and the inventiveness of human invention in preserving and enhancing the flavors of regional ingredients. In addition, pickling makes it possible to experiment with novel taste combinations and culinary techniques, which gives home cooks and chefs the opportunity to express their creative side in the kitchen.

In addition to its numerous gastronomic advantages, pickling also provides a number of practical benefits, including the ability to plan meals, store them, and make them more convenient. Due to the fact that pickled foods are simple to produce and require only a minimal amount of cooking skills or equipment, they are accessible to home cooks of all capabilities. Pickled foods may also be preserved for lengthy periods of time without the need for refrigeration, making them a great option for outdoor activities which includes picnics, camping vacations, and other activities that take place outside where refrigeration may be limited. They are a useful pantry staple for busy households and lifestyles that are constantly on the move because pickled foods are portable and may be stored for an extended period of time.

In conclusion, the advantages of pickling go well beyond the simple act of preserving food. Both home cooks and professional chefs can take advantage from the numerous benefits that pickled foods offer. These advantages include the enhancement of flavor and texture, the enhancement of nutritional content, and the promotion of culinary innovation. The art of pickling is a time-honored tradition that honors the art of preservation and the pleasure of tasting foods that are both distinctive and savory. This is despite the fact that we are constantly discovering new methods to appreciate and enjoy the wealth that nature provides.

Basic Equipment and Ingredients

Pickling is known as a time-honored method of food preservation that takes only a small amount of required equipment and supplies to get started. Despite the fact that the particular instruments and flavorings may change based on the recipe and the individual's preferences, there are a few fundamental components that are vital to the pickling process that are typically utilized. To become an expert in the art of pickling and to create delectable handmade pickled delicacies, it is essential to have a solid understanding of these fundamental components.

A non-reactive pot or pan, such as one made of stainless steel or enamel-coated cookware, is one of the most important pieces of equipment that an individual needs in order to successfully pickle food. Because they do not react with acidic components like vinegar, these materials are favored because they guarantee that the pickling solution will continue to be safe for consumption and will not impart any off-flavors to the final product. In addition, when it comes to preparing ingredients, a robust cutting board and a sharp knife are absolutely necessary. This is true whether you are slicing cucumbers for pickles or chopping onions for relish.

Glass jars or containers are another piece of equipment that is necessary for pickling because they are used to store the items that have been pickled. Glass jars are generally preferred over plastic or metal containers due to the fact that they are non-reactive and do not take on any of the aromas or scents that are produced by the pickling solution. When it comes to pickling, one of the most common containers used is a Mason jar with a screw-top lid. This type of jar offers an airtight seal, which helps to preserve the pickled items and extends their shelf life. Before using the jars, it is essential to make certain that they have been fully cleaned and sanitized in order to avoid contamination and damage to the contents.

The process of pickling involves a few fundamental materials in addition to the necessary equipment in order to produce the pickling solution and flavorings. Vinegar, which serves as a natural preservative and lends a sour flavor to the foods that are pickled, is the primary component in the majority of recipes for pickling. Vinegars which includes white vinegar, apple cider vinegar, and rice vinegar are frequently used in pickling, and each of these vinegars has a flavor profile that is distinctively its own. Water is frequently added to the pickling solution in addition to vinegar in order to produce the required level of tartness and to dilute the acidity of the pickling solution.

In the process of pickling, salt is an additional component that is vital since it helps to extract moisture from the materials and enhances the flavor of the items that are chosen to be pickled. Kosher salt, often known as pickling salt, is the salt of choice for pickling since it is simple to dissolve and does not include any additions or anti-caking agents that could potentially alter the clarity of the pickling solution. Sugar is another ingredient that is frequently used in pickling. Its purpose is to counteract the acidity of the vinegar and to impart a touch of sweetness to the final product. Certain extra flavorings, such as spices, herbs, and aromatics, may be added to the pickling solution in order to enhance the flavor and scent of the items that have been pickled. This is something that is determined by the recipe.

The pickling process begins with the preparation of the materials, which includes putting them in a pot or pan with the pickling solution. This is done after the fundamental equipment and ingredients have been constructed. The mixture is then brought to a boil, and after a few minutes, it is allowed to simmer for a few minutes so that the tastes may combine and the components can become slightly more absorbent. In the final step, the pickled foods are placed inside of glass jars that have been sterilized, then the pickling solution is poured over them, and the lids are then placed on top of them to make an airtight seal.

In conclusion, the equipment and ingredients that are necessary for pickling are straightforward and easily accessible, which makes it possible for home cooks of any ability to participate in the process. Everyone is capable of mastering the art of pickling and creating delectable handmade pickled delicacies that can be enjoyed throughout the year with just a few fundamental tools and pantry supplies necessary. It doesn't matter if you're looking for a traditional dill pickle, a spicy pickled pepper, or a tangy bread-and-butter pickle; the choices for pickling are virtually limitless; the only thing that can restrict you are your imagination and your culinary preferences.

CHAPTER III

The Art of Fermentation

What is Fermentation?

Microorganisms which includes bacteria, yeasts, and molds are responsible for the fermentation process, which is a natural metabolic activity that takes place in the absence of oxygen. During fermentation, carbohydrates and sugars are converted into alcohol, acids, and gases. This transforming process has been utilized by humans for thousands of years in order to generate a vast variety of fermented foods and beverages, ranging from bread and cheese to beer and wine. Even though fermentation is most commonly associated with the manufacturing of alcoholic drinks, it actually plays a considerably more significant role in the preservation of food, the generation of flavors, and the increase of nutritional value across a wide range of cultures and cuisines around the world.

In its most fundamental form, fermentation is propelled by the activity of microbes, which, through enzyme processes, convert more complicated sugars into molecules that are easier to understand. According to the particular fermentation process and the kinds of microorganisms that are involved, these microbes consume sugars and produce a variety of byproducts, including alcohol, lactic acid, acetic acid, and carbon dioxide. These byproducts are produced by the fermentation process. The metabolic activity of these bacteria not only helps to preserve the food, but it also improves its flavor, aroma, and texture. This results in the creation of distinctive and intricate profiles that are highly valued in culinary traditions all over the world.

Alcoholic fermentation is one of the most well-known types of fermentation. In this type of fermentation, yeasts use anaerobic respiration to convert carbohydrates into alcohol and carbon dioxide. Alcoholic beverages which includes beer, wine, and spirits are produced by this method, in which the sugars are sourced from grains, fruits, or other fermentable sources. This procedure is utilized to make these beverages. The type of yeast that is utilized, in addition to other elements such as temperature, pH, and the duration of the fermentation process, can have an effect on the flavor, aroma, and alcohol concentration of the final product. This can result in a diverse spectrum of alcoholic beverages that each have their own unique qualities.

Lactic acid fermentation is another popular type of fermentation that is used in the manufacture of food. This type of fermentation is utilized in addition to alcoholic fermentation. During the process of lactic acid fermentation, bacteria that produce lactic acid transform glucose into lactic acid. This process results in a decrease in the pH of the surrounding environment and produces a flavor profile that is characterized by a sour taste. This technique is utilized in the creation of fermented foods which includes yogurt, cheese, sauerkraut, and kimchi. Additionally, the acidic atmosphere not only helps to preserve the food, but it also gives a particular flavor and texture to the meal. In the process of making sourdough bread, wild yeasts and lactobacilli bacteria ferment the sugars in the flour, which results in a loaf that is tangy and tasty. Lactic acid fermentation is also involved in the manufacture of sourdough bread.

In addition to its function in the preservation of food and the development of flavor, fermentation also provides a number of nutritional advantages. Beneficial bacteria, also known as probiotics, are found in abundance in fermented meals. These bacteria have the ability to aid digestion and gut health. These probiotics are found in the intestinal tract, where they contribute to the preservation of a healthy balance of microorganisms and support the function of the immune system. Furthermore, fermentation has the potential to enhance the bioavailability of particular nutrients, so rendering them more readily available to the cells of the body. Fermentation, for instance, is responsible for the breakdown of phytic acid, which is found in grains and legumes. Phytic acid has the ability to impede the absorption of minerals such as iron as well as zinc, but fermentation makes these nutrients more easily available for human consumption.

A greater appreciation for artisanal and traditional food practices, as well as an enhanced understanding of the health advantages associated with fermented foods, have both contributed to a resurgence of interest in fermentation in recent years. This interest has been driven by a growing appreciation for fermentation. Home chefs and food aficionados alike are anxious to explore the countless possibilities that fermentation in the kitchen has to offer, and they are in the process of experimenting with fermented delicacies such as homemade yogurt, sourdough bread, kombucha, and other fermented beverages. In addition, a growing number of chefs and food manufacturers are mixing fermented products into their culinary creations, which results in dishes that have a greater depth of flavor and the ability to be more complex.

In conclusion, fermentation is a process that has been around for millennia and has been considered a natural and time-honored method. It has been integral to human culture and cuisine. Fermentation provides a multitude of benefits that continue to fascinate taste buds and nourish bodies all over the world. These benefits include the manufacturing of alcoholic drinks, the preservation of foods, and the development of nutritional content. We are honoring the transforming power of microbes and the ageless art of fermentation as we continue to explore the vast variety of fermented foods and beverages. This allows us to strengthen our connection to culinary traditions from the past as well as the present.

History of Fermentation

There is evidence that fermented foods and beverages date back thousands of years, indicating that the history of fermentation is as old as the history of human civilization itself to begin with. Fermentation is a natural metabolic process that is driven by microorganisms such as bacteria, yeasts, and molds. For millennia, humans have utilized fermentation as a means to preserve food, enhance flavor, and create a broad variety of culinary delights. Archaeological evidence reveals that early civilizations such as the Mesopotamians, Egyptians, and ancient Chinese were familiar with the process of fermentation and its transformational effects on food and drink. Although the actual origins of fermentation are shrouded in the mists of time, it is believed that these civilizations were aware of the process.

The creation of alcoholic beverages for consumption, such as beer and wine, was one of the oldest forms of fermentation that was done by ancient civilizations. Beginning in 4000 BCE, the Sumerians of Mesopotamia were the first people to create fermented beverages from barley and other grains. They are commonly attributed as being the inventors of beer. In a similar manner, the ancient Egyptians planted grapes and fermented them into wine, which was significant in their society for both religious and ceremonial purposes. The art of brewing and winemaking expanded throughout the ancient globe, carried by merchants, travelers, and conquerors. It eventually became an essential component of the cultural and social life of many different communities.

Additionally, fermentation was utilized in the production of a wide range of fermented foods, including bread, cheese, and pickles, in addition to the production of alcoholic beverages. Using sourdough starter, for instance, the ancient Egyptians cooked bread. Sourdough starter is defined as a mixture of wheat and water that has been colonized by wild yeast and bacteria. This mixture leavens the dough and gives the bread its distinctively sour flavor. It was common practice in ancient Rome to make cheese, and cheese factories known as "casei" produced a wide range of cheeses from milk that had been fermented with rennet or lactic acid bacteria. In a same manner, pickling, which is a type of fermentation, was utilized to preserve in vinegar or brine various types of vegetables, fruits, and meats in order to guarantee a consistent supply of food during times of scarcity.

Over the course of human history, fermentation has been an essential component in the development of culture and the preservation of human life. Monasteries in medieval Europe became centers of fermentation expertise. As part of their daily lives, monks experimented with brewing beer, making wine, and fermenting foods. Monasteries began to become hubs of fermentation knowledge. These monastic brewing traditions are responsible for the creation of some of the most well-known beer styles in the world, such as the Trappist ales of Belgium and the abbey beers of Germany. Similarly, cheesemaking flourished in monastic communities, with monks inventing distinctive cheese varieties that represented the local terroir and culinary traditions. This was especially true in the case of cheese.

There was a significant impact that fermentation had on international trade and exploration during the Age of Exploration, which occurred between the 15th and 16th centuries. During long sea voyages, when fresh food was rare and spoiling was a constant worry, European sailors relied on fermented foods such as sauerkraut, pickled herring, and hardtack biscuits to feed them. These foods were used to keep them alive. Not only did fermented foods supply important nutrients and calories, but they also assisted in the prevention of scurvy, which is a vitamin C shortage that is frequently experienced by sailors who go on extended excursions. The broad adoption of fermentation techniques by European explorers resulted in the facilitation of the interchange of culinary traditions between continents and contributed to the globalization of food and drink.

The process of fermentation has witnessed a resurgence in interest and appeal in more recent times. This resurgence has been spurred by a growing respect for artisanal and traditional food practices, as well as an improved understanding of the health advantages connected with fermented foods. People who enjoy cooking at home and people who are passionate about food are rediscovering the pleasures of fermented foods such as yoghurt, sourdough bread, kombucha, and other fermented treats. They are eager to explore the countless possibilities that fermentation poses in the kitchen. In addition, a growing number of chefs and food manufacturers are mixing fermented products into their culinary creations, which results in dishes that have a greater depth of flavor and the ability to be more complex.

In conclusion, the history of fermentation is a demonstration of the inventiveness and resourcefulness of human beings in harnessing the power of microorganisms to change basic resources into foods and beverages that are delectable and nutritious. Fermentation has been an essential component of human culture and gastronomy for thousands of years, beginning with the ancient civilizations of Mesopotamia and also Egypt and continuing through the monastic breweries of medieval Europe and the worldwide excursions of the Renaissance. We are honoring the transforming power of microbes and the ageless art of fermentation as we continue to explore the vast variety of fermented foods and beverages. This allows us to strengthen our connection to culinary traditions from the past as well as the present.

Health Benefits of Fermented Foods

Foods that have been fermented have been consumed for thousands of years, and they are highly valued for their distinctive aromas and textures, in addition to the potential health benefits they may offer. The process of fermentation is a natural process that is driven by helpful microorganisms such as bacteria, yeasts, and molds. These microorganisms convert carbohydrates and sugars into alcohol, acids, and gasses using fermentation. The food is not only preserved through this transformational process, but it also improves the nutritional worth of the food and promotes the health of the gut. As a result of this, fermented foods have garnered recognition for their role in facilitating digestion, enhancing immunity, and enhancing overall well-being.

The contribution that fermented foods make to the health of the gut is one of the most important health benefits that they offer. There are trillions of bacteria that live in the human digestive tract. These microbes are collectively called as the gut microbiota, and they play an important part in digestion, the absorption of nutrients, and the functioning of the immune system. The good bacteria that are found in fermented foods are referred to as probiotics, and they have the ability to assist in the replenishment and maintenance of a healthy balance of microorganisms in the gut. The consumption of fermented foods that are high in probiotics, which includes yogurt, kefir, kimchi, and sauerkraut, can encourage the growth of good bacteria in the gut, which in turn improves digestion and lowers the risk of digestive diseases like irritable bowel syndrome (or IBS) and inflammatory bowel disease (or IBD).

Fermented foods, in addition to positively impacting gut health, have the potential to strengthen the immune system and improve immunological function. Both the regulation of immune responses and the protection against infectious diseases are significantly influenced by the microbiota that lives in the gut. Those probiotics that are generated from fermented foods have the ability to help boost the immune system and lessen the risk of illnesses such as colds, influenza, and respiratory tract infections. They do this by modifying the makeup and activity of the microbiota that are found in the gut. The consumption of fermented foods regularly may also have anti-inflammatory benefits, which can help to minimize the symptoms of chronic inflammation and the health issues that are connected with it, such as arthritis, asthma, and autoimmune illnesses.

In addition, fermented foods have the potential to enhance nutritional absorption and bioavailability, which in turn makes it simpler for the body to absorb and make use of critical vitamins, minerals, and other nutrients. Through the process of fermentation, complex carbohydrates, proteins, and lipids are broken down into simpler molecules, which makes them easier to digest and absorb into the body through the gastrointestinal tract. Fermentation, for instance, has the potential to boost the availability of B vitamins like B12, folate, and riboflavin, in addition to minerals like iron, zinc, and magnesium. In addition, fermentation has the potential to increase the bioavailability of phytonutrients, antioxidants, and other bioactive chemicals that are present in plant-based diets, hence further boosting the nutritional value of these foods.

Furthermore, fermented foods are typically easier to digest than their equivalents that have not been fermented, which makes them acceptable for persons who have digestive sensitivities or intolerances. During the fermentation process, the food is partially predigested, which means that it is broken down into simpler forms that are easier for the body to consume. As a consequence of this, fermented foods may be better accepted by persons who suffer from illnesses such as lactose intolerance or gluten sensitivity. This enables these individuals to enjoy a greater variety of foods without experiencing discomfort or digestive distress. Additionally, fermented foods have the ability to help weight control and metabolic health, which is another significant health advantage of fermented foods. The consumption of fermented foods may be connected with a lower risk of metabolic syndrome, obesity, as well as type 2 diabetes, according to new study that has recently come to light. The consumption of fermented foods may be associated with the regulation of hunger, the reduction of cravings for unhealthy foods, and the improvement of insulin sensitivity, all of which contribute to the promotion of weight loss and metabolic health. In addition, the fermentation process has the potential to generate short- chain fatty acids (SCFAs), which have been demonstrated to have positive impacts on metabolism, energy expenditure, and fat storage.

In conclusion, fermented foods provide a wide range of health benefits, including the enhancement of nutrient absorption and metabolic health, as well as the promotion of gut health and immunity. Incorporating fermented foods that are rich in probiotics into one's diet, such as yogurt, kefir, kimchi, and sauerkraut, can assist in the maintenance of a healthy balance of gut bacteria, provide support for digestion, and lower the risk of digestive diseases and infections. In addition, fermented foods are abundant in important nutrients, phytonutrients, and antioxidants, which makes them valuable additions to a diet that is well-balanced. We are developing a deeper respect for the ancient art of fermentation and the role it plays in fostering well-being and vitality as we continue to investigate the possible health advantages of fermented foods.

Common Fermented Foods Around the World

In many different cultures around the world, fermented foods are considered to be an essential part of the culinary traditions. These foods are highly prized for their distinctive flavors, textures, and nutritional benefits. Fermented foods are a reflection of the inventiveness and creativity of human cultures in harnessing the power of bacteria to transform raw ingredients into culinary delights. Examples of fermented foods include the sour kimchi that is popular in Korea and the creamy yogurt that is popular in Greece. There are a number of common elements that emerge, which underline the universality of fermentation as a preservation technique and flavor-enhancing strategy. Although the specific types of fermented foods may differ from place to location, there are common motifs that show up.

The practice of fermentation is firmly ingrained in the culinary traditions of Asia, and the continent is home to a diverse range of fermented foods that are loved by its inhabitants. Kimchi is a fermented cabbage dish that began in Korea and is known for its spicy and acidic flavor. It is considered to be one of the most iconic fermented meals. A process known as lacto-fermentation is used to produce kimchi. This process involves the use of napa cabbage, radishes, and a spicy chili sauce known as gochujang. During this process, lactic acid bacteria convert carbohydrates into lactic acid, which is what gives the meal its distinctive lemony flavor. A staple in Korean cuisine, kimchi is a dish that is eaten as a side dish with practically every meal and is highly regarded for its multifaceted flavor profile as well as its numerous health advantages.

To a similar extent, fermentation plays a significant part in the traditional cuisine of Japan, with a variety of fermented dishes being consumed all over the country. Miso is a thick paste that is prepared from fermented soybeans, rice, or barley. It is frequently used as a condiment in meals like as soups, marinades, and sauces. Miso is one of the most common fermented foods. During the lengthy fermentation process that miso goes through, the koji mold, also known as Aspergillus oryzae, breaks down the proteins and carbohydrates that are found in the soybeans, resulting in a flavor that is rich in umami. Natto, which is another traditional Japanese fermented cuisine, is created from fermented soybeans with the addition of Bacillus subtilis bacterium. This bacteria is responsible for the dish's peculiar stringy texture and pungent aroma.

When we move to Europe, we find that fermented foods are likewise firmly established in the culinary traditions of the region, with each region boasting its own unique delicacies. Sauerkraut is a popular fermented cuisine that is produced in Eastern Europe. It is traditionally made with cabbage that has been finely shredded and then fermented with lactic acid bacteria. As a condiment, a side dish, or a component in classic dishes like pierogi and sausages, sauerkraut is a product that is enjoyed by culinary enthusiasts. Fermented dairy products, which includes yogurt, kefir, and sour cream, are considered to be essential components of the diet in Central Europe. These products are highly regarded for their velvety consistency and tangy flavor. Some people prefer to consume these fermented dairy products on their own, while others prefer to utilize them in cooking and baking to enhance the richness and depth of flavor of their dishes.

There is also a significant amount of fermentation that takes place in the Mediterranean region, with olives being one of the most well-known fermented foods. Olives are subjected to a procedure known as brine fermentation, in which they are submerged in a solution of saltwater in order to eliminate bitterness and improve culinary quality. You can eat fermented olives as a snack, an appetizer, or as a component in salads, sandwiches, and other foods that are typical of the Mediterranean region. Cheese is another fermented food that is commonly consumed in the Mediterranean region. Each country in the region has its own distinctive varieties of cheese as well as its own production methods. Fermented cheeses all across the world, from feta cheese in Greece to Parmesan cheese in

Italy, are highly regarded for the intricate flavors and textures that they possess.

The consumption of fermented foods, such as injera, kisra, and ogi, is widespread across the African continent. Fermentation is an important component of many traditional diets in Africa. Injera is a type of fermented flatbread that is produced from teff flour, which is indigenous to Ethiopia and Eritrea and does not contain gluten. Because of its characteristic sour flavor and spongy texture, injera is the ideal companion to stews and curries. This is because the fermentation process gives injera its characteristic flavor. Another fermented flatbread that is popular in Sudan and other East African nations is called kisra. Kisra is produced from sorghum or millet flour and is fermented. Additionally referred to as pap or akamu, ogi is a fermented cornmeal porridge that is popular in West Africa. It is typically consumed as a staple morning food, and it can be topped with milk, sugar, or savory ingredients.

Chicha, a fermented maize beer that is popular in Peru and other Andean countries, is one of the many traditional foods and beverages that are produced through the process of fermentation in South America. Chicha is produced by fermented maize (corn) that is chewed and spat out by women who are referred to as "chichaeras." This procedure brings about the enzymatic breakdown of starches into sugars resulting in the production of chicha. After the chewed maize has been cooked, filtered, and fermented with wild yeast, the end product is a beverage that has a flavor that is slightly sour and has a moderate amount of alcohol. Sauerkraut, which is also known as curtido in Central America, is another fermented meal that is popular in South America. It is created from fermented cabbage, carrots, and onions, and it is frequently used as a condiment with pupusas and other traditional dishes.

In conclusion, fermented foods are an essential part of the culinary traditions of countries all over the world. These foods are highly prized for their distinctive flavors, textures, and health related advantages. Fermented foods are a reflection of the diversity and ingenuity of human cultures. From the spicy kimchi of Korea to the creamy yogurt of Greece, examples of fermented foods demonstrate how human cultures have used the power of bacteria to convert raw ingredients into gastronomic delights. As we continue to investigate the diverse array of fermented foods that can be found all over the world, our admiration for the age-old art of fermentation and the fundamental role it plays in the formation of our culinary history and cultural identity grows ever more profound.

CHAPTER IV

Setting Up Your Pickling and Fermentation Kitchen

Essential Equipment

Before beginning the process of pickling and fermenting, it is necessary to acquire certain key pieces of equipment in order to guarantee excellent outcomes and results that are delightful. It doesn't matter if you're fermenting cabbage to make sour sauerkraut or pickling cucumbers to make crunchy dill pickles; having the appropriate equipment at your disposal can make the process go more smoothly and efficiently. Glass jars or fermentation crocks are vital vessels for housing the ingredients and allowing the fermentation process to take place. These vessels are among the fundamental pieces of equipment that are required for pickling and fermentation. Materials that are safe for food, such as glass or ceramic, should be used to construct these containers. These materials are non-reactive and will not impart any flavors that are not desired to the foods that are getting fermented.

In addition to containers, a dependable pair of kitchen scales is necessary for correctly measuring materials and ensuring that the appropriate proportions are maintained in recipes involving pickling and fermentation. In the process of fermentation, precision is essential since even minute differences in the proportions of the ingredients can have a significant effect on the fermentation process and the flavor of the fermented meals. Furthermore, a decent pair of kitchen scales may assist assure uniformity and reproducibility in your experiments with pickling and fermentation. This will enable you to fine-tune your recipes and obtain the best possible outcomes.

Another piece of equipment that is important for pickling and fermentation is a fermentation weight or tamper that is made of food-grade materials. This piece of equipment helps to ensure that the components remain submerged in the brine or liquid during the fermentation process. For the purpose of preventing mold and ensuring even fermentation, it is essential to keep the ingredients submerged. However, exposing the ingredients to oxygen might result in the development of off-flavors and spoiling. There is a wide range of shapes as well as sizes available for fermentation weights, which can be anything from disks made of glass or ceramic to inserts made of silicone or plastic that are designed to fit within the fermentation tank. However you decide to go about it, you need to make sure that it fits snugly inside the container and that it offers sufficient pressure to ensure that the components remain submerged.

In addition, a fermentation airlock or lid is necessary for the creation of an anaerobic environment within the fermentation vessel. This environment allows carbon dioxide to depart while preventing oxygen from entering the vessel. This serves to avoid the growth of germs that could be detrimental to the organism and guarantees that the fermentation process is successful. Fermentation airlocks are available in a variety of forms, such as water-sealed airlocks and one-way valve lids. These airlocks all serve the same objective, which is to enable gas to escape while still maintaining a sealed environment. Whatever type you decide to go with, you need to make sure that it can be attached to your fermentation tank in a secure manner and that it offers a tight seal to prevent air from getting in.

When it comes to monitoring the acidity of the foods you are fermenting and making sure they reach the right level of tanginess and flavor, a pH meter or pH strips can be extremely helpful tools. It is possible to adapt the fermentation process according to the requirements of the situation by using pH meters, which are more accurate than pH strips and provide readings of the acidity level in real time. pH strips, on the other hand, are more reasonably priced and may still provide a rough estimate of the acidity level; hence, they are ideal for use in home fermentation projects. By keeping an eye on the pH of the items you are fermenting, you may help prevent over-fermentation or spoiling, which will ultimately result in a product that is both safe and delicious.

Last but not least, if you have a fermentation station or area in your kitchen that is specifically designated for fermentation, you may simplify the process of pickling and fermentation, as well as maintain order and efficiency. It is possible that there will be a specific countertop or shelf that you will use to store your fermentation vessels, equipment, and supplies. Additionally, it will provide you with a location to routinely monitor and check on the foods that are fermenting. When it comes to preventing cross-contamination with other foods and minimizing the danger of deterioration or contamination during the fermentation process, having a production station that is specifically designated for fermentation can be of great assistance. Additionally, it allows you to experiment with a variety of recipes and methods, as well as keep track of many fermentation projects at the same time.

In conclusion, having the appropriate equipment is absolutely necessary for the accomplishment of effective pickling and fermentation operations. Equipment such as glass jars or fermentation crocks, kitchen scales, fermentation weights, airlocks or lids, pH meters or strips, and a fermentation station that is specifically designed for fermentation are some of the important equipment that are required to ensure a fermentation process that is both smooth and enjoyable. You will be able to embark on your fermentation adventure with confidence and inventiveness if you make an investment in quality equipment and set up a dedicated fermentation space in your kitchen. This will allow you to uncover the delicious flavors as well as health benefits of fermented foods that you have manufactured at home, such as pickles, krauts, and other fermented foods.

Safety and Hygiene Practices

When it comes to pickling and fermentation, safety and cleanliness are of the utmost importance in order to guarantee the manufacture of fermented foods that are not only flavorful but also delicious and nutritious. In spite of the fact that pickling and fermentation are natural processes that have been utilized for the preservation of food for millennia, it is vital to maintain good sanitation and hygiene procedures in order to avoid contamination and food spoiling. It is possible for home cooks to enjoy the process of pickling and fermentation while simultaneously insuring the safety and quality of their homemade creations if they adhere to a few important rules.

In the first place, it is absolutely necessary to begin the process of pickling and fermenting foods using utensils and equipment that have been thoroughly cleaned and disinfected. Jars made of glass, fermentation crocks, cutting boards, knives, and any other implements that are utilized during the process are included in this category. It is possible to eliminate any dirt, debris, or germs that can contaminate the fermenting foods by washing the equipment and utensils with hot soapy water and completely rinsing them before using them. Sanitizing the equipment using a solution of diluted bleach or a sanitizer that is safe for use with food will further reduce the chance of infection and guarantee that the fermentation environment is clean.

In addition, in order to get the greatest results when pickling and fermenting foods, it is essential to utilize ingredients that are both fresh and of a specific quality. Before pickling or fermenting, fresh fruits and vegetables should be washed thoroughly under running water to eliminate any dirt or pesticides that may be present. In addition, it is vital to use salt that has not been iodized, such as kosher salt or pickling salt, because iodine has the potential to impede the growth of microorganisms that are useful throughout the fermentation process. In addition, the utilization of water that has been filtered or is free of chlorine can be of assistance in preventing the fermentation process from being influenced by undesirable chemicals or additives.

Additionally, in order to guarantee a satisfactory conclusion, it is vital to keep the temperature and humidity levels maintained at the appropriate levels during the fermentation process. Fermentation is most successful when it takes place within a particular temperature range, which is normally between 60 to 75 degrees Fahrenheit or 15 to 24 degrees Celsius, however this temperature range might change based on the kind of fermentation and the microorganisms that are involved. Fermentation can be slowed down or stopped entirely by temperatures that are too high, while temperatures that are too high can encourage the growth of germs that are toxic to the food and cause it to go bad. In a similar vein, ensuring that the humidity level is maintained at the appropriate level, which is often somewhere between 70 and 80 percent, helps to produce an ideal environment for fermentation and prevents the growth of mold or yeast.

One other essential component of safety and hygiene in pickling and fermentation is the management of the fermentation environment in order to forestall the development of pathogens and bacteria that are hazardous to the organisms being fermented. To do this, it is necessary to keep fermenting foods submerged in brine or liquid in order to generate an anaerobic environment that prevents the growth of bacteria that are dependent on oxygen production. The utilization of fermentation weights, airlocks, or one-way valve lids can be of assistance in accomplishing this goal. These devices allow gasses to exit while simultaneously preventing air from entering the fermentation vessel. In addition, it is possible to prevent contamination and guarantee the security of the final product by periodically monitoring the fermentation process and eliminating any batches that smell off or have mold on them.

Additionally, in order to avoid the spread of contamination from one food to another, it is necessary to maintain a high level of personal hygiene when handling and fermenting food. This includes cleaning one's hands thoroughly with soap and water before and after handling fermenting foods, as well as avoiding touching one's mouth, nose, or eyes while dealing with fermenting components. Additionally, it is important to avoid contacting the eyes, nose, and mouth. Additionally, it is recommended to wear clean kitchen attire, such as aprons or gloves, in order to prevent the spread of bacteria from clothing to food whilst working in the kitchen. To further maintain the safety of the fermentation process, it is important to minimize cross- contamination with other foods and surfaces, such as raw meats or vegetables that have not been cooked. This can help hinder the spread of bacteria that are hazardous to the fermentation process.

Finally, it is vital to carefully store fermented foods after fermentation has taken place in order to preserve their quality and ensure their safety. Foods that have been fermented should be kept in clean, sealed containers and kept in the refrigerator or in a cool, dark place to avoid the fermentation process from accelerating and to prevent the food from going bad. Fermented foods that are stored correctly can be preserved for a number of months or even years, despite the fact that they may continue to ferment and develop new flavors during the course of consumption. It is essential to label fermented foods with the date that they were prepared, and it is also essential to dispose of any batches that exhibit symptoms of decomposition or contamination, such as mold, scents that are not pleasant, or textures that are slimy.

In conclusion, the implementation of safety and hygienic measures is crucial in order to guarantee the success of pickling and fermentation projects, as well as the manufacture of fermented foods that are not only safe but also pleasant and healthy. Pickling and fermentation are two processes that home cooks can enjoy while minimizing the risk of contamination and spoilage. This can be accomplished by following proper sanitation procedures, using fresh ingredients, maintaining optimal fermentation conditions, controlling the fermentation environment, practicing good personal hygiene, and storing fermented foods in the appropriate manner. Pickles, krauts, and other fermented foods that are created at home can be a delectable and nutritious addition to any diet if they are prepared with a moderate amount of attention to detail and care.

Organizing Your Workspace

It is absolutely necessary to have a workspace that is well-organized in order to achieve success with pickling and fermenting projects. Whether you are fermenting cabbage to make sauerkraut or pickling cucumbers to make crunchy dill pickles, having a workstation that is clean, efficient, and organized will help you speed the process and ensure that you get consistent results. Through the implementation of a few essential ideas, home cooks can set up their fermentation station to achieve optimum production while also maximizing their enjoyment.

Initially and most importantly, you should dedicate a particular place in your kitchen or home for the purpose of fermenting and pickling substances. You may use a countertop, table, or shelf to create a space in your kitchen where you can arrange your fermentation vessels, equipment, and supplies without causing any more clutter in other sections of the kitchen. If you have a fermentation station that is specifically designed for fermentation, you will be able to store all of your pickling and fermentation supplies in a one location. This will make it much simpler for you to get everything you require and will reduce the likelihood of cross-contamination with other foods.

The next step is to organize your fermentation station by putting objects that are similar together in groups and ensuring that tools and supplies that are regularly used are within easy reach. On one side of your workspace, for instance, you should keep your fermentation vessels, such as glass jars or crocks, together. On the other side of your workspace, you should store your fermentation weights, airlocks, and lids. You should put your cutting boards, knives, measuring cups, and any other tools in a spot that is easily accessible so that you can quickly get them whenever you require them. When the fermentation process is being carried out, this not only helps save time and work, but it also helps avoid disruptions and distractions from occurring.

Additionally, in order to maintain the orderliness and cleanliness of your fermentation supplies, you need make an investment in storage options such as containers, baskets, or shelves. When you are creating recipes for pickling and fermenting, it is helpful to store materials such as salt, vinegar, spices, and other flavorings in containers that are clearly labeled. This will make it easier for you to locate the specific items you require. It is important to keep your fermentation vessels and equipment clean and organized, and to store them in a specific location when they are not being used. This will help you avoid clutter and keep your workspace organized.

Maintaining adequate sanitation and hygiene practices is another essential component of managing your fermentation station. This is done to minimize contamination and spoiling of the product through improper handling. Before and after each usage, make sure that your fermentation containers and equipment are thoroughly cleaned and sterilized. To remove any dirt, debris, or germs, you should use hot, soapy water and sanitizers that are approved for use on food. Before handling fermenting foods, make sure to fully wash your hands with soap and water. Additionally, prevent yourself from contacting your face or any other surfaces while you are dealing with fermenting components. Also, make sure that your fermentation station is cleaned and sanitized on a regular basis in order to prevent the accumulation of mold, yeast, or any other potentially hazardous microbes.

Additionally, in order to maximize both productivity and ease, you should take into consideration the architecture and flow of your fermentation station. Make sure that your workspace is organized in a way that makes it simple for you to switch between different duties, such as preparing the materials, filling the fermentation vessels, and monitoring the fermentation process. Keep supplies and tools that are frequently used within easy reach, and reduce the amount of clutter and obstructions that aren't required and could potentially slow down your production. Not only does this assist save time and energy, but it also lessens the likelihood of mishaps or spills occurring while the fermentation process is taking place.

If you want to establish a comfortable and pleasant working atmosphere, the last thing you should do is make sure that your fermentation station has adequate ventilation and lighting. When it comes to selecting and fermentation projects, natural light is perfect since it enables you to see the colors and textures of your ingredients more clearly and gives an ambiance that is welcoming for cooking and experimenting. If there is a lack of natural light in your workstation, you might want to think about installing overhead fixtures or task lighting to illuminate your workspace and make it simpler to see what you are doing. In addition, make sure that there is sufficient ventilation to eliminate any extra heat, steam, or aromas that are produced during the fermentation process. This will facilitate the maintenance of a clean and comfortable working environment.

In conclusion, it is vital to organize your workspace for pickling and fermenting operations in order to guarantee productivity, uniformity, and safety. It is possible for home cooks to create a fermentation station that is both functional and enjoyable by designating a particular area for fermentation, grouping items that are similar together, investing in storage solutions, maintaining proper sanitation and hygiene practices, optimizing layout and flow, and ensuring adequate lighting and ventilation. You will be able to embark on your adventure of pickling and fermentation with self-assurance and inventiveness if you have a workspace that is well-organized. This will allow you to uncover the delicious flavors and health advantages of fermented foods that you have created at home.

CHAPTER V

Getting Started with Pickling

Pickling Techniques and Methods

The process of pickling is a time-honored method of preservation that has been utilized for millennia to increase the shelf life of fruits and vegetables, as well as to create delectable sauces and accompaniments. Pickling techniques and procedures vary greatly from culture to culture and cuisine to cuisine, reflecting the diversity and originality of pickled foods all around the world. These techniques and methods include relishes, chutneys, and pickles that have a sour flavor. Although the fundamental principles of pickling remain the same, which include submerging items in a brine or an acidic solution in order to prevent the growth of bacteria that cause spoiling, there are a number of various techniques to pickling, each of which has its own distinctive qualities and flavor profiles.

Vinegar pickling is one of the most prevalent pickling procedures. In this method, fruits or vegetables are submerged in a solution consisting of vinegar, water, salt, and sugar, along with various spices and flavorings including garlic, dill, and mustard seeds. Because vinegar pickling produces pickles that are tart and acidic, with a crisp texture and lively taste, it is an excellent method for pickling vegetables that are sturdy, such as cucumbers, onions, peppers, and other vegetables. The acidity of the vinegar assist in preserving the pickles and impedes the growth of harmful germs, which ensures that the product is safe to consume and may be stored for an extended period of time. In addition to being consumed directly from the jar, vinegar pickles can also be utilized as a condiment or component in the preparation of sandwiches, salads, and appetizers.

The fermentation method is another popular pickling technique. In this approach, fruits or vegetables are submerged in a brine made of saltwater and allowed to ferment naturally over a period of time while being preserved. By utilizing the power of beneficial bacteria, often known as lactic acid bacteria, fermentation pickling is able to convert carbohydrates into lactic acid, resulting in a flavor profile that is nuanced and sour. There are a variety of fermented pickles that are appreciated all over the world, including sauerkraut, kimchi, and lacto- fermented cucumbers. These pickles are desirable because of the probiotics that they contain and their distinctive flavor. Pickles that are fermented go through a transformational process that softens the texture of the vegetables and creates a sour flavor. As a result, they are an adaptable and healthful addition to any meal.

A further point to consider is that rapid pickling, which is sometimes referred to as refrigerator pickling or fresh-pack pickling, is a straightforward and adaptable form of pickling that does not involve any heat processing or prolonged fermentation. In order to make quick pickles, fruits or vegetables are first submerged in a brine that is based on vinegar, along with spices and flavorings. After this, the pickles are placed in the refrigerator for a brief length of time so that the flavors can combine. As a result of their ability to preserve the crisp texture as well as fresh flavor of the raw materials, quick pickles are an excellent choice for vegetables that are delicate in nature, such as carrots, asparagus, and cucumbers. Home cooks have the ability to experiment with different combinations and produce pickled creations that are both distinctive and delectable by using quick pickles, which can be modified with a broad variety of spices and flavorings according to their preferences.

In addition, brine pickling is a classic method of pickling that is utilized to preserve fruits and vegetables in a solution of salt for a longer period of time. In brine pickling, the product is submerged in a brine that is composed of water, salt, and occasionally sugar or other flavorings. After this, the brine is let to sit for a period of time that can range from several weeks to several months so that the brine can ferment and develop taste. Brine pickles are renowned for their robust taste and crisp texture. There are numerous types of brine pickles, including kosher dill pickles and bread & butter pickles, which are extensively used in American cuisine. Brine pickling is a labor of love for people who are passionate about pickles since it involves careful attention to the concentration of salt and the amount of time that fermentation takes in order to create the proper flavor and texture.

In conclusion, pickling techniques and processes provide a wide variety of possibilities for preserving fruits and vegetables as well as adding taste to them. It doesn't matter if you're looking for the tangy acidity of vinegar pickles, the probiotic health of fermented pickles, the crisp crunch of fast pickles, or the powerful flavor of brine pickles; there is a pickling process that can accommodate every taste and preference. Home cooks have the opportunity to unleash their creativity and explore the unlimited possibilities of pickled foods by experimenting with various pickling procedures. This allows them to add zest and flavor to their meals while also preserving the crop for pleasure throughout the year. Everyone is capable of mastering the art of pickling and making wonderful homemade pickles that are both a treat for the taste buds and a source of nourishment for the body if they put in a little bit of time and effort.

Basic Pickling Recipes (e.g., Pickled Vegetables, Fruits)

With the help of pickling, home cooks are able to transform fresh vegetables into tangy, savory delicacies that can be enjoyed throughout the year. Pickling is a sort of culinary art. Recipes for basic pickling give a varied and tasty method to improve your culinary inventions. Whether you are preserving the wealth of the garden or simply adding a punch of flavor to your meals, simple pickling recipes offer a way to elevate your culinary innovations. There are a variety of pickling recipes available to cater to an array of preferences and tastes, including fiery pickled peppers, sweet pickled fruits, and crisp pickled cucumbers.

The humble cucumber is often considered to be one of the most well-known and well-liked pickled vegetables. Pickled cucumbers, which are often referred to as pickles, are a staple ingredient in a wide variety of cuisines all over the world. They are crisp and refreshing. Pickled cucumbers are normally prepared by soaking them in a brine that is composed of vinegar, water, salt, sugar, and various spices including dill, garlic, and mustard seeds. Standard pickled cucumber recipes utilize this process. The acidity of the vinegar assist in preserving the cucumbers and gives them their distinctive tangy flavor. The spices offer depth and complexity to the dish, while the vinegar helps to preserve the cucumbers. Cucumbers that have been pickled can be consumed as a snack or an appetizer directly from the jar. They can also be used to add crunch and character to dishes such as relishes, salads, and sandwiches.

In a similar vein, pickled peppers are another popular option for home picklers who are trying to add some heat to their meals. You may tailor pickled peppers to your taste and level of tolerance for heat, whether you prefer gentle banana peppers, sour jalapenos, or spicy habaneros. Pickled peppers can be personalized to meet your preferences. Typical pickled pepper recipes require for the peppers to be sliced and then packed into jars with a brine that is based on vinegar. Garlic, onions, and many other flavorings are also used in the recipe. It is possible to use pickled peppers to give pizzas, sandwiches, and tacos a kick. Additionally, pickled peppers can be cut up and tossed into dips, salsas, and marinades to offer additional flavor and heat.

Additionally, pickled onions are a tasty and flexible complement to any kitchen's arsenal of vegetables and herbs. Pickled onions, which are tangy and somewhat sweet, offer a splash of color and acidity to a variety of dishes, including salads, sandwiches, tacos, and more. Recipes for pickled onions typically involve slicing red or white onions thinly and soaking them in a brine that is produced from vinegar, water, sugar, and salt. Additionally, the brine may contain spices such as peppercorns, bay leaves, and coriander seeds. Onions that have been pickled can be consumed directly from the jar as a condiment or garnish, or they can be utilized to impart a burst of flavor to an array of recipes.

The next topic we will discuss is pickled fruits, which are a lovely addition to any meal or snack because of their sweet and sour flavor. Pickled fruits, whether they are peaches, pears, or plums, give a burst of flavor and acidity that goes well with sweet and also savory dishes. Pickled fruits go well with both types of dishes. Simmering the fruit in a syrup that is prepared from vinegar, water, sugar, and other spices like cinnamon, cloves, and star anise is the approach that is commonly taken in the preparation of basic pickled fruit recipes. The sweetness of the fruit counteracts the sourness of the vinegar, resulting in a flavor profile that is harmonic and may be described as both refreshing and satisfying. Eat pickled fruits on their own as a dessert or snack, or use them to add a sweet and tangy touch to salads, cheese platters, and drinks. Pickled fruits can be enjoyed in any of these ways.

In conclusion, basic pickling recipes provide home cooks with a world of taste and inventiveness, allowing them to preserve and enrich the wealth that the garden has to give. Irrespective of whether you are pickling fruits, veggies, or a combination of the two, there is a pickling recipe that may accommodate any circumstance or preference. When it comes to pickling, the chances are virtually limitless. Some examples include pickled cucumbers that are crisp, pickled peppers that are spicy, and pickled fruits that are sweet. Mastering the art of pickling and enjoying the delectable pleasures of handmade pickled delicacies is something that can be accomplished by anyone with just a few basic materials and a little bit of imagination.

Troubleshooting Common Pickling Problems

Home cooks have the ability to transform fresh food into sour and savory pleasures through the use of pickling, which is a time-honored method of cooking preservation. On the other hand, just like any other culinary venture, pickling can potentially face obstacles at various points along the way. In order to achieve good outcomes and delicious results, it is important to recognize and fix typical pickling problems. These problems might range from hazy brine to mushy pickles.

Mushy pickles are a problem that many people who pickle their own food at home may experience. There are a number of factors that might lead to mushy pickles, including overripe or improperly picked cucumbers, incorrect handling and processing processes, so forth. It is vital to select firm, fresh cucumbers that do not exhibit any symptoms of bruising or soft places in order to avoid pickles that are on the mushy side. In addition, make sure to remove the blossom end of the cucumber before pickling it, as this part of the cucumber contains enzymes that can cause the softening process to occur. In addition, in order to prevent the pickles from becoming mushy, it is important to avoid overprocessing or overcooking them. Instead, pay close attention to the directions provided in the recipe and keep a close eye on the pickles as they are being processed to ensure that they maintain their crispness and firmness.

In addition, murky brine is a common problem that arises during the pickling process. This can be detrimental to the appearance and flavor of the vegetables that are pickled. In most cases, cloudy brine is the result of an excessive amount of minerals or contaminants in the water that was used to generate the brine, in addition to an inadequate amount of rinsing of the vegetables prior to the pickling process. Use filtered or distilled water while manufacturing brine to reduce the amount of minerals and other contaminants in the water. This will prevent the brine from becoming hazy. In addition, when you pickle the vegetables, make sure to give them a thorough washing under running water to remove any dirt and/or debris that can cloud the brine. Cloudy brine is generally safe to drink, but it may alter the appearance of the pickles. If cloudy brine does occur, it is safe to ingest.

In addition, pickles that are slimy or soft can be an indication that they have gone bad or that the fermentation process was not done correctly. Pickles that are slimy or soft might be the consequence of numerous factors, including contamination with microorganisms that cause deterioration, insufficient salt concentration, or insufficient fermentation time. Make sure that all of the equipment and utensils are clean and sanitized before use, and make sure that you use the appropriate amount of salt concentration and fermentation time that is mentioned in the recipe. This will prevent the pickles from becoming slimy or squishy. Be sure to keep the pickles submerged in the brine throughout the fermentation process. This will prevent the pickles from being exposed to oxygen, which can promote the development of bacteria that are damaging to the pickles.

Mould growth on the surface of brine or pickles is another cause for concern because it may be an indication that the brine or pickles have gone bad or that they have been contaminated. Mold can grow on pickles if they are not properly submerged in brine or if the fermentation tank is not adequately sealed. Both of these factors might potentially lead to mold growth. If you want to avoid mold from growing on the pickles, you should make sure that the fermentation vessel is tightly shut and that fermentation weights or other means are used to keep the pickles immersed in the brine. Additionally, prevent air from entering the fermentation vessel. In the event that mold formation does take place, it is recommended that the contaminated pickles be thrown away and the fermenting vessel be cleaned before beginning a fresh batch.

Furthermore, uneven processing or inadequate brine penetration might be the cause of varying flavor or texture in pickled vegetables. This can occur when the veggies get pickled. Be sure to cut the veggies into uniform sizes and pack them snugly into the jars before adding the brine. This will guarantee that the flavor and texture of the vegetables remain constant throughout the process. In addition, make sure that the pickles are allowed to ferment or process for the appropriate amount of time as mentioned in the recipe. This will guarantee that the pickles develop the flavor and texture that you wish. If inconsistencies continue to occur, you might want to think about modifying the recipe or the method in order to attain the desired outcomes.

In conclusion, in order to solve typical pickling issues, it is necessary to pay close attention to the details and have a solid understanding of the pickling process. Home picklers have the ability to ensure successful outcomes and excellent results by addressing difficulties such as mushy pickles, hazy brine, soft or slimy pickles, mold growth, and variable flavor or texture results. Everyone is capable of mastering the skill of pickling and enjoying the savory pleasures of handmade pickled treats if they are patient, practice, and willing to explore a little bit.

CHAPTER VI

Exploring Fermentation

Fermentation Processes and Methods

Fermentation is a natural metabolic process that takes place when microorganisms like bacteria, yeasts, or molds digest sugars and carbohydrates in the absence of oxygen. This process results in the production of a variety of byproducts, including alcohol, acids, and gases. Food has been preserved, its flavor has been enhanced, and its nutritional content has been improved through the application of this time-honored method for thousands of years. Methods and procedures of fermentation can vary greatly based on a number of aspects, including the type of microorganisms that are involved, the substrate that is being fermented, and the product that is wanted at the end of the process. In order to properly harness the power of microbes in the kitchen, it is essential to have a solid understanding of the various fermentation processes and methodologies. These include anything from wild fermentation to controlled fermentation and include everything in between.

Wild fermentation is one of the most traditional and extensively employed ways of fermentation. In this method, the fermentation process is initiated by naturally occurring microorganisms that are present in the environment or on the surface of the food. In order to inoculate the fermentation vessel and get the fermentation process started, wild fermentation makes use of the natural diversity of microorganisms that are found in the air, soil, and water, as well as on the surfaces of fruits and vegetables. The foods sauerkraut, kimchi, sourdough bread, and kombucha are all examples of foods that have been fermented in the wild. Unpredictability and variability are two characteristics that are commonly associated with wild fermentation. This is due to the fact that the precise strains of microbes that are involved might change based on conditions such as temperature, humidity, and time.

Controlled fermentation, on the other hand, is a method of fermentation that is more structured and controlled. In this method, particular strains of microorganisms are put into the fermentation vessel in order to commence and direct the fermentation process. Controlled fermentation is a technique that is frequently utilized in the commercial food production and fermentation sectors, which place a high priority on maintaining uniformity, quality, and safety. When producers choose and cultivate particular strains of bacteria, yeasts, or molds, they are able to guarantee that the fermentation process will proceed in a predictable manner and result in the production of the product that they desire. The production of an array of fermented foods as well as beverages, such as cheese, yogurt, beer, wine, and vinegar, is accomplished through the process of controlled fermentation.

Anaerobic fermentation, on the other hand, is a type of fermentation that takes place in the absence of oxygen. This sort of fermentation normally takes place in an anaerobic environment, such as a fermentation vessel that is sealed or submerged in a liquid. Foods and beverages that have been fermented through the process of anaerobic fermentation include lacto-fermented vegetables, pickles, and sauerkraut, among other fermented foods and beverages. Lactic acid bacteria are accountable for the production of lactic acid being a byproduct of anaerobic fermentation. This process takes place in the absence of oxygen and involves the reaction of sugars and carbohydrates. As a result of this process, an acidic environment is produced, which prevents the growth of germs that cause food to go bad and helps to preserve the food.

Additionally, aerobic fermentation is a sort of fermentation that takes place in the presence of oxygen, often in an aerobic environment such as the surface of a liquid or solid substrate. Anaerobic fermentation is the type of fermentation that takes place inside the absence of oxygen. Acetic acid microorganisms metabolize ethanol in the presence of oxygen during the process of aerobic fermentation, which leads to the production of acetic acid as a byproduct. This process also results in the production of fermented foods such as vinegar and kombucha. The sour and acidic flavor that is characteristic of vinegar and kombucha is produced as a result of this process. The process of aerobic fermentation necessitates the utilization of aeration and agitation in order to guarantee that the microorganisms receive an appropriate amount of oxygen and to facilitate the fermentation process.

Furthermore, spontaneous fermentation is a type of fermentation that takes place without the intentional addition of microorganisms. Instead, it relies on the natural microflora that is present in the environment or on the surface of the food to commence the fermentation process. Traditional fermented foods and beverages, such as sourdough bread, beer, and wine, are typically produced by the process of spontaneous fermentation. This method is also commonly utilized. In the process of spontaneous fermentation, the naturally occurring microorganisms that are present in the environment inoculate the fermentation vessel and direct the fermentation process, which ultimately results in flavors that are both distinctive and complex.

In conclusion, fermentation techniques and procedures play an important role in the creation of fermented foods and beverages, since they are responsible for molding the flavor, texture, and nutritional content of these products. The art of fermentation provides a plethora of opportunities for culinary innovation and discovery, regardless of whether it is accomplished by spontaneous fermentation, controlled fermentation, anaerobic fermentation, aerobic fermentation, or wild fermentation. By gaining an understanding of the many processes and methods of fermentation, home fermenters are able to access the transforming potential of microbes and experience the delectable pleasures of fermented treats that they have manufactured themselves.

Fermentation Vessels and Tools

Fermentation vessels and equipment are crucial components of the fermentation process. They provide a regulated environment in which microorganisms may perform their enchantment and change raw ingredients into fermented foods and beverages that are delectable. There are a broad variety of alternatives accessible to home fermenters, ranging from classic crocks and jars to sophisticated fermenting kits and equipment. These solutions are made to meet the requirements and preferences of home fermenters. It is possible for home fermenters to select the appropriate equipment for their fermentation projects and obtain excellent results if they have a thorough understanding of the many types of fermentation containers and vessels used for fermentation.

The fermentation crock is a large ceramic container that has a water-sealed cover that allows gasses to escape while preventing air from entering the vessel. It is considered to be one of the most classic and flexible fermentation vessels. Fermentation crocks are an excellent choice for fermenting huge quantities of vegetables, fruits, or beverages like sauerkraut, kimchi, or kombucha. The lid that is sealed with water produces an anaerobic atmosphere inside the crock, which helps to promote efficient fermentation and prevents the food from going bad. An additional benefit of the ceramic material is that it helps manage temperature and humidity, which results in a more consistent environment for fermentation.

Glass jars are another common type of vessel used for fermentation. These jars are available in a wide range of sizes and shapes, and they are ideal for fermenting small to medium-sized batches of beverages including vegetables, fruits, or beverages. Home fermenters are able to monitor the fermentation process and detect any changes in color, texture, or aroma since glass jars are transparent that allow them to see what is happening. In addition, glass jars are non-reactive and simple to clean, which makes them an additional advantageous choice for fermentation projects that are carried out at home. In order to ferment within glass jars, all that is required is to pack the components into the jar, cover it with a lid or an airlock designed for fermentation, and then wait for the fermentation process to take place.

Furthermore, fermentation airlocks are useful instruments for producing an anaerobic environment inside fermentation vessels. This atmosphere allows gases to exit while preventing air from entering the vessel. Fermentation airlocks are available in a variety of forms, such as water-sealed airlocks, one-way valve lids, and silicone gaskets. These airlocks all serve the same objective, which is to enable gas to escape while still maintaining a sealed environment. For the purpose of preventing mold growth, spoiling, and off-flavors, the utilization of a fermentation airlock is beneficial since it establishes a regulated fermentation environment within the vessel. Crocks, jars, and fermenting buckets are only some of the fermentation vessels that are compatible with fermentation airlocks. Also suitable are fermenting buckets.

Additionally, fermentation weights or tamper are crucial tools for maintaining the submersion of components in the brine or liquid during the fermentation process. As a result of the fact that exposure to oxygen can result in spoiling and off-flavors, it is essential to keep components submerged in order to avoid the formation of mold and to ensure that fermentation occurs evenly. There is a wide range of shapes as well as sizes available for fermentation weights, which can be anything from disks made of glass or ceramic to inserts made of silicone or plastic that are designed to fit within the fermentation tank. Fermentation tampers are used in a similar manner to press down on the ingredients and expel any air bubbles that have become trapped. This ensures that the components are packed tightly and produce an even fermentation.

As an additional point of interest, pH meters or pH strips are extremely helpful instruments for measuring the acidity of fermenting foods and beverages and ensuring that they achieve the correct level of tanginess and flavor. Home fermenters are able to make adjustments to the fermentation process as required because to the real-time readings that pH meters provide regarding the acidity level. Because pH strips are more reasonably priced and still have the ability to provide a reasonable estimate of the acidity level, they are excellent for undertaking fermentation projects at home. Making ensuring that the pH of fermenting foods is monitored helps to prevent over-fermentation or rotting, and it also guarantees that the final result is both safe and appetizing.

In conclusion, fermentation vessels and instruments are indispensable for the accomplishment of successful fermentation projects. They offer a regulated setting in which microorganisms can perform their enchantment and change raw components into delectable fermented foods and beverages. Home fermenters have a broad variety of alternatives available to them to choose from in order to meet their requirements and preferences, whether they choose to use conventional fermenting crocks and jars or contemporary fermenting kits and equipment. Home fermenters can commence on their fermentation adventure with confidence and inventiveness if they choose the appropriate fermentation vessels and tools and acquire a grasp of their functions. This will allow them to uncover the delectable flavors and health advantages that are associated with fermented pleasures that they have concocted themselves.

Introduction to Starter Cultures

Starter cultures are extremely important in the process of fermentation because they act as catalysts that trigger the fermentation process and direct its progression. It is the responsibility of these living microorganisms, which are often bacteria or yeasts, to convert the sugars and carbohydrates present in the substrate into a variety of byproducts, including acids, alcohols, and gases. These byproducts contribute to the flavor, texture, and nutritional profile of the final product. In some fermentations, the inoculation of the fermentation vessel is accomplished through the use of wild or naturally occurring microorganisms. However, commercially available starter cultures are utilized in other fermentations to guarantee consistency, predictability, and safety. The importance of understanding the role that starting cultures play in the fermentation process cannot be overstated for home fermenters who want to achieve good outcomes and delicious products.

Lactic acid bacteria are known as one of the most prevalent starter cultures utilized in the process of fermentation. These bacteria are accountable for the production of lactic acid, which is a consequence of the fermentation process. Fermentation of dairy products like yogurt, cheese, and kefir, as well as fermented vegetables which includes sauerkraut and kimchi, generally involves the employment of lactic acid bacteria, which are present naturally on the surfaces of fruits and vegetables. These bacteria are also commonly used in the fermentation process. These bacteria are able to grow in more acidic settings and are well-suited to the low pH circumstances that are present during fermentation. They are able to outcompete dangerous bacteria and contribute to the preservation and safety of the end product within these conditions. In addition, during the fermentation process, lactic acid bacteria produce a variety of taste compounds, which are responsible for the characteristic sour and complex flavors that are associated with fermented foods.

Yeast is another form of starting culture that is frequently used in fermentation. During the fermentation process, yeast is responsible for turning carbohydrates into alcohol and carbon dioxide. It is usual practice to use yeasts in the fermentation process of alcoholic beverages like beer, wine, and cider, as well as bread and other baked foods. Yeasts can be found naturally on the surfaces of fruits and grains, as well as in the environment. Yeasts are extremely adaptable microorganisms that can ferment an array of substrates and produce a wide range of taste compounds. The fermentation process is dependent on a number of important parameters, including temperature, pH, and the availability of nutrients. In addition, yeasts are essential components in the leavening process of bread and other baked goods, which results in the characteristic rise and texture of these baked goods.

Additionally, acetic acid bacteria are a form of starter culture that is used in the process of fermentation. These bacteria are accountable for the production of acetic acid as a byproduct of the fermentation process. Bacteria that produce acetic acid are frequently utilized in the fermentation process of vinegar. In this process, with the presence of oxygen, these bacteria transform ethanol into acetic acid. The sour and acidic flavor that is distinctive of vinegar is produced by this procedure, which also contributes to the preservation of the finished product. Acetic acid bacteria are highly aerobic microorganisms that require oxygen to flourish. Because of this, they are ideally suited to the surface of liquid substrates such as wine or cider that are undergoing fermentation to produce vinegar.

In addition, for the purpose of inoculating the substrate and directing the fermentation process, mold cultures are occasionally utilized in the fermentation process. Koji is a type of mold that is used to manufacture enzymes that break down starches and proteins into sugars and amino acids. These enzymes are then fermented by other microorganisms like as yeasts or bacteria. Mold cultures are often employed in the fermentation process of koji. Koji is an essential component in the manufacturing of an array of fermented foods as well as beverages, including soy sauce, miso, and sake, where it plays a role in the development of flavors and the efficiency of fermentation. In conclusion, starter cultures are necessary for initiating and directing the fermentation process. They supply the living microbes that are required to convert sugars and carbohydrates into a variety of byproducts, including acids, alcohols, and gases. Whether they are lactic acid bacteria, yeast, acetic acid bacteria, or mold cultures, starter cultures are extremely important in the process of fermenting foods and beverages because they significantly influence the flavor, texture, as well as nutritional profile of the final product. Home fermenters have the ability to achieve successful outcomes and enjoy the tasty rewards of fermented pleasures that they have manufactured themselves if they have a thorough understanding of the role that starter cultures play in the fermentation process and if they choose the appropriate cultures for their fermentation projects.

Basic Fermented Food Recipes (e.g., Sauerkraut, Kimchi, Kombucha)

For millennia, fermented foods have been an indispensable component of diets all over the world. They are highly regarded for their singular flavors, numerous nutritional advantages, and extended shelf life. Basic fermented food recipes provide home fermenters with a wide variety of options that are both delicious and nutritious. These recipes range from tangy sauerkraut to spicy kimchi and effervescent kombucha. The power of helpful microbes is harnessed in these time-honored recipes, which turn raw ingredients into dishes and beverages that are rich in flavor and probiotics.

The fermented dish known as sauerkraut is renowned for its acidic flavor and crisp texture, making it one of the most well-known and favorite fermented foods. Sauerkraut is a mainstay in Eastern European cuisine and a versatile component that can be used in a broad variety of meals. It is created by fermenting shredded cabbage with salt and sometimes the addition of other flavorings like caraway seeds or juniper berries. To make sauerkraut, simply put shredded cabbage with salt in a fermentation vessel, such as a crock or glass jar, and pack it tightly to release the juices. This will allow the cabbage to become sauerkraut. The fermentation process of the cabbage should be slowed down by placing it in the refrigerator after it has been allowed to ferment at room temperature for a period of time ranging from several days to several weeks, all depending on the level of sourness that you desire. As a result, a condiment that is both sour and crunchy is produced. This condiment can be consumed on its own or added to enhance the flavor and crunch of foods such as sausages, salads, and sandwiches.

Kimchi, on the other hand, is a traditional Korean fermented food that is created from vegetables like cabbage, radishes, and scallions, and it is seasoned with chili paste, garlic, ginger, and fish sauce. The spicy and pungent flavor of kimchi, along with its rich scent and crunchy texture, are some of the reasons why it is so popular. To begin the process of making kimchi, you must first prepare the veggies by seasoning them with salt and then allowing them to wilt so that their juices can be released. The next step is to combine the chili paste, garlic, ginger, and fish sauce with the veggies, making sure that the seasoning is evenly distributed throughout the vegetables. A fermentation vessel, such as a crock or a glass jar, should be used to pack the veggies that have been seasoned. The vegetables should then be allowed to ferment at room temperature for a period of time ranging from several days to several weeks, depending on the level of fermentation that you prefer. Rice, noodles, and grilled meats are all excellent complements to the hot and sour condiment that is the end result of this process.

Furthermore, kombucha is a fermented tea beverage that has been increasingly popular in recent years due to the fact that it has a flavor that is both refreshing and perhaps beneficial to one's health. The fermentation of sweetened tea with a symbiotic culture of bacteria and yeast (SCOBY) results in the production of kombucha, which is highly regarded for its bubbliness and sourness respectively. Beginning with a batch of sweetened tea made from black, green, or herbal tea leaves, you will need to begin the process of making kombucha. Once the tea has reached room temperature, move it to a fermentation jar along with the SCOBY and some starting liquid from a previous batch of kombucha. Allow the tea to ferment once it has reached room temperature and before it is consumed. You should let the kombucha ferment at room temperature for a period of time that can range anywhere from a few days to weeks, depending upon the level of acidity and effervescence that you prefer. Cover the vessel with a cloth or a lid that allows air to circulate while avoiding contamination. The end product is a beverage that is slightly carbonated and has a sour flavor. It can be consumed on its own or flavored with fruits, herbs, and/or spices to add more layers of layers of complexity.

In conclusion, basic fermented food recipes such as sauerkraut, kimchi, and kombucha offer a delicious and nutritious method to introduce fermented foods into your diet. These recipes can have a positive impact on your health. Whether you enjoy the sour crunch of sauerkraut, the spicy kick of kimchi, or the refreshing fizz of kombucha, these traditional recipes are simple to create at home and offer an infinite number of possibilities for experimenting with different flavors. Home fermenters have the ability to unlock the delectable flavors and health benefits of fermented delicacies that they have manufactured at home by harnessing the power of helpful microbes from the environment.

CHAPTER VII

Advanced Techniques in Pickling and Fermentation

Experimenting with Flavors and Ingredients

It is possible for people to change fresh produce into dishes that are palatable and can be stored for an extended period of time by using ancient preservation techniques such as pickling and fermentation. These techniques have been employed for millennia. The possibility to experiment with a broad variety of flavors and ingredients is one of the pleasures that come with pickling and fermentation. These flavors and ingredients can range from traditional combinations to combinations that are daring and unexpected. It doesn't matter if you're fermenting fruits, pickling vegetables, or making beverages that are high in probiotics; the possibilities for flavor exploration are virtually limitless; the only thing that can restrict you are your imagination and your personal preferences in terms of flavor.

When it comes to pickling, the process begins with the selection of fresh produce and the formulation of a tasty brine or marinade. From there, the experimenting with flavors and ingredients can begin. In spite of the fact that conventional pickling recipes typically call for components like vinegar, salt, and sugar, there is a great deal of potential for creativity and personalization. Experimenting with different kinds of vinegar, such as apple cider vinegar, rice vinegar, or even red wine vinegar, can result in the creation of dishes that have flavors that are both distinctive and intricate. A similar approach can be taken to add depth and complexity to pickled vegetables by altering the sorts of spices and herbs that are used in the brine and the amounts that are used. The choices for taste combinations are practically limitless, and they range from dill and garlic to coriander and peppercorns or everything in between. Incorporating sweeteners like honey or maple syrup into the pickles is another way to add a touch of sweetness to the pickles while also balancing out the acidity of the vinegar.

In a similar vein, fermentation provides a variety of chances for flavor experimentation. This is because the fermentation process has the ability to enhance and modify the flavors of the substances. When it comes to fermentation, the process of experimentation starts with the selection of fresh produce and the establishment of an atmosphere that is conducive to fermentation. While salt is an essential component in the majority of fermentation recipes, the kinds of salt and the amounts of salt that are used might vary based on the individual's preferences in terms of flavor and the outcomes that are wanted. In addition, the flavor profile of fermented foods and beverages can be altered by experimenting with various kinds of sugars, such as honey, molasses, or fruit juice instead of sugar. For instance, adding fruit juice to a beverage that is fermenting, such as kombucha, might result in the introduction of new flavors and fragrances, in addition to giving more nutrients for the microorganisms that are carrying out the fermentation process.

In addition, you may take pickling and fermentation to new heights of culinary inventiveness by experimenting with different flavorings and incorporating other additions. When it comes to pickling, the addition of aromatic components like ginger, lemongrass, or star anise can make the pickles taste more exotic and fragrant than they would otherwise. In a similar vein, experimenting with various kinds of vegetables, such as carrots, beets, or radishes, can add color and texture to the pickled masterpieces that you make. Through the process of fermentation, the addition of flavorings like herbs, spices, or botanicals can result in the creation of complex and varied flavors in fermented foods and beverages. For instance, adding juniper berries to sauerkraut that is fermenting can give a delicate piney aroma and earthy flavor. Similarly, adding fresh herbs like thyme or rosemary to fermented veggies can add a fresh and herbal note to the vegetables.

Furthermore, the flavor and texture of fermented and pickled foods can be influenced by experimenting with different fermentation vessels and using different fermentation processes. For instance, fermenting vegetables in a traditional clay crock or ceramic fermentation pot can lend a distinctive earthy flavor to the finished product. On the other hand, fermenting vegetables in glass jars or stainless-steel containers may generate a flavor profile that is cleaner and more neutral. In a similar vein, home fermenters have the ability to customize the flavor and texture of their creations to suit their preferences by experimenting with different fermentation durations and temperatures. Longer fermentation times often result in more intense flavors and softer textures, whilst shorter fermentation times typically result in flavors that are less intense and textures that are crunchier.

In conclusion, the process of pickling and fermenting provides an infinite number of options for culinary innovation and exploration. These involve experimenting with different flavors and kinds of ingredients. The freedom to modify flavors and ingredients enables home fermenters to produce one-of-a-kind and delectable concoctions that are a reflection of their individual preferences and tastes. This is true whether the fermenting process involves pickling vegetables, fermenting fruits, or making beverages that are high in probiotics. Home fermenters have the ability to unlock the full potential of pickling and fermentation by embracing experimentation and allowing their creativity to guide them. This allows them to experience the wonderful rewards of handmade pickled and fermented delicacies.

Utilizing Seasonal Produce

In order to preserve the abundance of seasonal vegetables, pickling and fermenting are two excellent methods that may be utilized. These methods capture the flavors of summer and allow them to be enjoyed throughout the winter months. Pickled and fermented meals that celebrate the abundance of each season can be made by home fermenters by utilizing the natural flavors and textures of fresh fruits and vegetables that are at their greatest maturity. These foods can be delicious and nutritious. By incorporating seasonal produce in pickling and fermentation, home fermenters are able to connect with the rhythms of nature and enjoy the freshest flavors throughout the year. This includes using crisp cucumbers and juicy tomatoes in the summer, as well as robust root vegetables and citrus fruits in the winter.

Pickling and fermentation provide a fantastic opportunity to preserve the harvest as well as enjoy the flavors of summer long after the season has passed. This is especially true during the summer months, when gardens and farmer's markets are bursting at the seams with an abundance of fresh vegetables. Some examples of seasonal treasures that can be transformed into delectable pickles, relishes, and fermented veggies include crisp cucumbers, juicy tomatoes, crunchy green beans, and aromatic herbs. These are just a few examples. Home fermenters have the ability to capture the brilliant colors and flavors of the season and enjoy them throughout the year by pickling or fermenting summer food when it is at its peak freshness.

A new assortment of seasonal food becomes available in the fall, as the temperature start to drop and the number of days begins to shorten. This new assortment of fruit offers a distinct palette of flavors and textures that can be used for pickling and fermentation process. During the fall months, there is an abundance of hearty root vegetables such as carrots, beets, and turnips, as well as crisp apples and pears. These veggies lend themselves wonderfully to pickling and fermentation. Spices like cinnamon, cloves, and nutmeg lend a sense of coziness and depth to ferments and pickles that are inspired by the fall season. Sweeteners like maple syrup or honey compliment the natural sweetness of autumn vegetables. Pickling or fermenting fall produce gives home fermenters the opportunity to relish the tastes of the season and enjoy a taste of autumn even after the leaves have fallen from the trees this time of year.

If you want to preserve the last of the harvest of the season and brighten up the colder months with bursts of taste and color, pickling and fermenting are two methods that you may use with the arrival of winter and the subsequent decrease in the availability of fresh fruit. During the winter months, citrus fruits like lemons, oranges, and grapefruits are in season. These oranges, lemons, and grapefruits can be transformed into tart citrus pickles, marmalades, or fermented beverages like citrus kombucha. During the winter months, there is an abundance of hardy vegetables such as cabbage, kale, and Brussels sprouts. These veggies are ideal for fermenting into crunchy sauerkraut or kimchi. The use of seasonal vegetables in pickling and fermenting allows home fermenters to experience the flavors of winter and maintain the flavors of the season even when fresh produce is limited. This allows them to keep the flavors of spring alive.

In the spring, when the temperature begins to warm up and new vegetation starts to develop in gardens and fields, pickling and fermentation provide an opportunity to enjoy the arrival of fresh greens and fruits that are in season at the beginning of your growing season. Because they are abundant in the spring, tender asparagus, crisp radishes, and colorful greens like spinach and arugula are ideal for pickling or fermenting into savory sauces, salads, or side dishes. These vegetables are also wonderful for making pickled pickles. To pickles and ferments that are inspired by spring, fresh herbs like dill, parsley, and chives offer a sense of brightness and freshness. Additionally, edible flowers like nasturtiums or violets can bring a touch of elegance and also beauty to fermented creations. For home fermenters, the ability to embrace the flavors of spring and have a taste of the season's abundance as it unfolds is made possible through the utilization of seasonal food in pickling and fermenting processes.

By incorporating seasonal vegetables in pickling and fermentation, home fermenters are able to relish the flavors that are associated with each season and enjoy the freshest ingredients throughout the entire year. In the summer, crisp cucumbers and juicy tomatoes are available, and in the winter, substantial root vegetables and citrus fruits are available. Seasonal produce provides a plethora of choices for pickling and fermenting. Home fermenters are able to connect with the rhythms of nature, relish the flavors of the harvest, and enjoy delicious and nutritious pickled and fermented foods throughout the year because they preserve the bounty of each season through the processes of pickling and fermentation.

Long-Term Storage and Preservation Methods

People have been using pickling and fermenting for a very long time as techniques of preserving food for an extended period of time. This has enabled societies to keep seasonal produce and enjoy it throughout the year. Utilizing the power of helpful microbes, these time-honored procedures transform fresh fruits and vegetables into foods that are not only delectable but also shelf-stable, allowing them to be enjoyed for months or even years after they have been prepared. There are a variety of long-term storage and preservation options accessible to home fermenters who are interested in extending the life of their pickled and fermented creations. These methods range from conventional methods such as canning and cellar storage to modern approaches such as vacuum sealing and refrigeration.

Canning is one of the most traditional techniques of long-term preservation in pickling and fermentation. Canning involves sealing pickled or fermented foods in sterile jars and then processing them in a boiling water bath or pressure canner to establish a vacuum seal. Canning is one of the most common means of long-term storage. The process of canning an item eliminates any bacteria that may still be present in the food and provides an airtight seal that protects the product from going bad or becoming contaminated. The fact that fermented foods like pickles, sauerkraut, and other fermented foods can be kept at room temperature for months or even years makes them an excellent choice for stocking the pantry with and enjoying throughout the year. Canned food is a tried-and-true way of preserving food for an extended period of time, despite the fact that it takes a certain amount of equipment and careful attention to detail.

The cellar storage method is another traditional way of long-term preservation in the pickling and fermentation processes. This approach involves preserving pickled or fermented foods in a cellar or root cellar that is cool, dark, and where temperatures remain consistent throughout the year. Storage in the cellar is dependent on the natural coolness and humidity of the cellar environment in order to slow down the fermentation process and prevent the fermentation from going bad. It is possible to keep pickled vegetables like sauerkraut, kimchi, and pickles in the basement for a number of months or even years if they are well sealed and protected from light and pests. Crocks and jars are the most common storage containers for pickled vegetables. The preservation of food for an extended period of time through the use of cellar storage is a low-tech and environmentally friendly strategy that has been utilized by communities all over the world for generations.

Furthermore, vacuum sealing is a contemporary method of long-term preservation that includes eliminating air from the packaging of pickled or fermented foods and then sealing them in airtight bags or containers. This approach is used to store foods for an extended period of time. By avoiding oxidation and reducing the growth of germs that cause food to go bad, vacuum sealing helps to retain the freshness and flavor of foods that have been fermented or pickled. There are a variety of fermented foods, including pickles, sauerkraut, and other fermented foods, that may be preserved in the refrigerator or freezer for extended periods of time. This enables home fermenters to enjoy their creations at the peak of their freshness. There is a widespread adoption of vacuum sealing as a method of long-term storage since it is both practical and efficient. This approach is utilized in both commercial food production and household kitchens.

In addition, chilling is a straightforward and efficient means of storing pickled and fermented foods for an extended period of time, particularly for examples of foods that have not been subjected to heat processing or vacuum sealing. Pickles, sauerkraut, and kimchi are examples of fermented vegetables that can be preserved in the refrigerator for several weeks or even months if they are stored in airtight containers. This will prevent the veggies from drying out and absorbing aromas. Both pickled and fermented foods benefit from refrigeration because it slows down the fermentation process and limits the growth of bacteria that cause deterioration. This helps to keep the texture and flavor of the foods. Refrigeration is a technique of long-term storage for home fermenters that is both practical and easily accessible, despite the fact that it requires regular maintenance and also requires electricity.

Pickled and fermented foods, particularly those that are not appropriate for canning or refrigeration, can also be stored for an extended period of time by freezing them. This is an additional alternative. Pickles, sauerkraut, and chutneys are examples of fermented fruits and vegetables that can be preserved in the freezer for a number of months or even years if they are portioned out into freezer-safe containers or bags. By effectively stopping the fermentation process and preserving the texture and flavor of fermented and pickled foods, freezing them is a fantastic choice for storing them for an extended period of time. Freezing is a convenient and versatile technique of long-term storage for home fermenters, despite the fact that it requires sufficient freezer space and may alter the texture of certain foods.

In conclusion, home fermenters who want to extend the life of their creations and enjoy the flavors of the harvest throughout the year have a variety of alternatives available to them through the use of long-term storage and preservation methods in pickling and fermentation. It is possible to preserve the wealth of the garden and enjoy the flavors of pickled and fermented foods for a considerable amount of time after the season has ended. This can be accomplished by an array of methods, including old-fashioned techniques like canning and cellar preservation, as well as more contemporary techniques like vacuum sealing and refrigeration. Those who ferment their own food at home can store their pantry with pickled and fermented delicacies that are both delicious and nutritious if they learn the technique of long-term storage and preservation. They can then enjoy these delights throughout the year.

CHAPTER VIII

Incorporating Pickled and Fermented Foods into Your Diet

Health Benefits and Nutritional Value

It has been known for a long time that pickled and fermented foods are highly valued for their distinctive flavors and their contribution to gut health; nevertheless, the advantages of these foods go far beyond taste alone. Foods that have been preserved using these ancient ways not only improve their flavor and extend their shelf life, but they also enrich them with essential nutrients and probiotics that are useful to the health and well-being of the individual as a whole. There is a wide array of health advantages that can be gained from include fermented and pickled foods in one's diet. These benefits include improved digestion and immunological function, enhanced nutrient absorption, and reduced inflammation.

Pickled and fermented foods have a high probiotic content, which is one of the primary reasons why they are beneficial to one's health. The fermentation process results in the proliferation of helpful bacteria such as lactobacilli and bifidobacteria, which then generate lactic acid. Lactic acid serves as a natural preservative and gives the meals a tangy flavor. These probiotic bacteria contribute to the maintenance of a healthy balance of the microbiota in the gut, which is one of the most important factors in digestion, immunological function, and overall health. Consuming foods that are high in probiotics, such as sauerkraut, kimchi, and yogurt, can assist in the replenishment and diversification of the microbiota in the gut, thereby enhancing digestion, enhancing immunity, and lowering the risk of gastrointestinal illnesses which includes irritable bowel syndrome (or IBS) and inflammatory bowel disease (or IBD).

In addition, foods that have been fermented or pickled are abundant in vitamins, minerals, and antioxidants. This is because the natural fermentation process and the retention of nutrients during fermentation both prevent the loss of nutrients. The vitamins K and vitamin C, as well as the minerals potassium, calcium, and magnesium, can be found in plenty in fermented vegetables like sauerkraut, pickles, and kimchi. These foods are especially rich in these nutrients. The immune system, bone health, as well as cardiovascular health are only some of the functions that are supported by these nutrients, which play an important part in maintaining these activities. Yogurt and kefir are two examples of fermented dairy products that are rich in protein, calcium, as well as B vitamins. These nutrients make them ideal additions to a diet that is intended to be balanced.

In addition, foods that have been pickled or fermented can aid enhance digestion and the body's ability to absorb nutrients by encouraging the growth of good bacteria in the lower gastrointestinal tract. For the purpose of making dietary fibers and complex carbohydrates more digestive and accessible, the probiotic bacteria that are present in fermented foods contribute to the breakdown and fermentation of these substances. This process not only helps improve digestion and reduce symptoms of digestive diseases such as bloating and constipation, but it also promotes the absorption of nutrients which includes vitamins, minerals, and phytonutrients that are found in the foods that we consume. Consuming foods that have been pickled or fermented on a regular basis can assist to maximize nutrient absorption and create a healthy microbiota in the gut, which subsequently supports overall health and well-being.

Additionally, foods that have been pickled or fermented have been connected to a variety of additional health benefits, including a reduction in inflammation, an improvement in immune function, and an improvement in mental health. Fermented foods contain probiotic bacteria that help control the immune system and reduce inflammation in the gut. This has been related to a lower risk of chronic inflammatory illnesses such as arthritis, allergies, and autoimmune diseases. Probiotic bacteria can be found in fermented meals. Additionally, there is a growing body of evidence that suggests the gut microbiota may have a role in the regulation of mood and mental health. Specifically, there are studies that have established a connection between dysbiosis (imbalance) in the gut microbiota and illnesses such as anxiety and depression. Consuming foods that are high in probiotics, which includes yogurt, kefir, and fermented vegetables, may assist to support a healthy gut-brain axis and enhance mental well-being.

In conclusion, foods that have been pickled or fermented offer a wide variety of health benefits and nutritional value, which makes them essential additions to a diet that is balanced. Incorporating these traditional foods into your diet can help support digestion, enhance immunity, and promote overall health and well-being. Some examples of these foods include sauerkraut and kimchi, which are rich in probiotics, and yogurt and kefir, which are filled with vitamins. Consuming foods that have been pickled and fermented on a regular basis as part of a diet that is both varied and balanced allows you to harness the power of beneficial bacteria and nutrients to fuel your body and enhance your overall health.

Recipes for Using Pickles and Ferments in Everyday Cooking

When it comes to the vast and varied world of culinary arts, pickles and ferments retain a distinct place. They provide a one-of-a-kind combination of tanginess, umami, and frequently a pleasant crunch, which can transform ordinary cuisine into something remarkable. When you utilize fermented foods and pickles in your dishes, you are not only able to add more complex flavors and textures, but you are also able to incorporate the health advantages that are connected with these foods, such as improved digestion and enhanced gut flora. The purpose of this section is to investigate the several ways in which pickles and ferments can be included into regular meals, so offering a canvas for both conventional and unique approaches to the preparation of food.

A common starting point for those who are interested in cooking with ferments and pickles is the straightforward process of introducing these ingredients into dishes that are already familiar to them. For a long time, sandwiches have been a canvas for pickles, which give a zesty bite that contrasts with the softness of bread and the richness of meats and cheeses. One illustration of this is their use in sandwich sandwiches. Pickles, on the other hand, have a wider range of use in the kitchen than only as a condiment may suggest. The vinegary kick that they give to salad dressings is a great way to brighten up grain and green salads. They can be cut very finely and added to salad dressings. Pickle juice, on the other hand, can be used as a marinade, which not only tenderizes meat but also imparts taste to it. This is an example of how the by-products of pickling can be just as beneficial as the pickles themselves.

The use of fermented foods in everyday cooking opens up a wide range of possibilities due to the profound flavors and probiotic properties that they possess. Not only may sauerkraut be used as a side dish, but it can also be utilized as a primary element in meals such as stews and casseroles, where it lends depth and tanginess. Sauerkraut's sour complexity makes it a versatile ingredient. In a similar vein, the fiery Korean ferment known as kimchi can be added to foods such as scrambled eggs, fried rice, or even pancakes, so imparting its signature flavor and crunch to dishes that would otherwise be considered straightforward. Due to the fact that fermented foods are so versatile, they may be changed to suit a broad variety of culinary traditions and individual preferences, ranging from the subtle to the bold.

When it comes to cooking with ferments and pickles, one of the most interesting elements is the numerous opportunities for creativity and originality that are presented. The use of pickled vegetables, for instance, can be pureed and used to make sauces or relishes. This provides a speedy method for adding depth to grilled meats or vegetables that have been roasted. A savory umami flavor that enriches the entire taste profile of a food can be achieved by incorporating fermented bean pastes, such as miso and tempeh, into soups, marinades, and dips. These pastes contribute to the overall flavor profile of the dish. At-home chefs have the opportunity to explore and produce their own distinctive culinary expressions by using these items, which have the ability to elevate the mundane to the spectacular.

Ferments and pickles, in addition to contributing to the enhancement of flavor, also contribute to the nutritional content of the meals they are used in. One process in particular, known as lacto-fermentation, is responsible for increasing the bioavailability of vitamins and minerals in foods, so rendering them more nutrient-dense. Furthermore, the probiotics that are commonly found in fermented foods are helpful to the health of the digestive tract, making them an appetizing approach to improve general well-being. People are able to enjoy meals that are not only delectable but also good to their health if they incorporate pickles and ferments into their regular cooking routine.

The cultural significance of using ferments and pickles in the kitchen is not something that can be ignored. There is a connection to legacy and history that can be found in many of these dishes because they have strong roots in certain cuisines and traditions. For the purpose of paying homage to historical traditions, cooks can incorporate pickles and ferments into their dishes, while also altering them to match the preferences and lifestyles of modern people. This combination of the traditional and the contemporary contributes to the enhancement of the culinary scene by fostering a more profound awareness for the myriad of ways in which food can be conserved, cooked, and enjoyed.

In addition, in this day and age, where there is a growing emphasis on being environmentally responsible, it is essential to take into mind the impact that cooking with ferments and pickles might have on the environment. These preservation techniques lengthen the shelf life of perishable foods, which in turn allows for a more effective utilization of resources, hence reducing the amount of food that is wasted. It is possible for individuals to make a contribution to a more sustainable food system by including ferments and pickles into their regular cooking routines. This type of food system prioritizes the reduction of waste while simultaneously maximizing flavor and nutrition.

To summarise, the incorporation of pickles and ferments into regular cooking presents a plethora of opportunities for enhanced flavour, enhanced nutritional value, and enhanced creative potential. Through the enhancement of flavors, textures, and health benefits, these ingredients have the potential to transform ordinary recipes into dinners that are unforgettable. The use of pickles and ferments in the kitchen offers up a whole new world of culinary possibilities, whether it be via the straightforward addition of pickles to a sandwich or through the creative application of fermented sauces in some dishes. Furthermore, the practice of adding these foods into regular meals is not only about appreciating their flavor, but also about establishing a connection with cultural traditions, encouraging health, and supporting methods that are sustainable in the food industry. The ancient methods of food preservation, such as pickling and fermenting, are being given a new lease on life as more people become aware of the pleasures of using them in the kitchen. These methods continue to enhance our tables and our lives with their particular flavors and the benefits they offer.

CHAPTER IX

Building a Resilient Food Supply

Importance of Food Preservation in Emergency Preparedness

It is impossible to overestimate the significance of effectively preserving food when it comes to emergency preparedness. It is of the utmost importance to have a secure food supply in this day and age, when conditions such as natural disasters, worldwide pandemics, and other unanticipated events have the potential to disturb the fragile fabric of daily living. When it comes to developing resilience, ensuring food security, and maintaining nutrition during times of emergency, the practice of food preservation is absolutely necessary. Throughout this section, the varied function that food preservation plays in emergency preparedness is investigated, with a particular emphasis placed on the value of food preservation in sustaining individuals and communities through difficult times.

Canning, freezing, drying, and fermenting are some of the methods that are included in the spectrum of food preservation techniques. These methods are used to extend the shelf life of food products. By inhibiting the development of bacteria that cause food to go bad and the action of enzymes that cause food to deteriorate, these approaches ensure that food will continue to be safe, nutritious, and tasty for longer periods of time. In the context of emergency preparedness, the ability to properly preserve food means that individuals and communities are able to maintain a steady food supply, even in situations where access to fresh produce is restricted or disrupted.

One of the most important aspects of being prepared for a disaster is having a strategic stockpile of preserved foods. There are a variety of food options that are long-lasting, space-efficient, and can be preserved for longer periods of time. These include freeze-dried meals, dried fruits and vegetables, and canned items. These foods that have been preserved provide the vital nutrients and calories that are required to maintain life, particularly in situations where traditional food supply systems undergo disruptions. The variety of preservation techniques also makes it possible to consume a wide array of foods, which is essential for preserving morale and ensuring adequate nutrition throughout extended periods of emergency.

Additionally, the method of food preservation is an essential component in the process of cultivating independence and self-sufficiency. It can be dangerous to rely on outside assistance for food during times of crisis because there is a possibility that supply lines will be disrupted or delayed. Learning how to properly preserve food gives individuals and communities the capacity to take charge of their own food security and empowers them to take care of themselves. This ability to rely on oneself is extremely crucial in times of crisis, when the ability to acquire and prepare food can be the deciding factor in whether or not one survives as opposed to enduring suffering.

Preservation of food is another factor that contributes to the effective utilization of resources. By increasing the shelf life of food, preservation techniques reduce the amount of food that is wasted and make the most of the resources that are available for food. In the event of an emergency, where resources need to be carefully managed to ensure their sustainability over possibly extended periods of time, this efficiency is of utmost importance. Foods that have been preserved do not need to be refrigerated, which reduces the need for energy and fuel, both of which may be in short supply during times of crisis. This particular component of food preservation is in line with broader sustainability and preparedness aims, and it encourages appropriate resource management in both normal and emergency situations.

There are substantial psychological ramifications associated with food preservation in the context of emergency preparedness, in addition to the practical benefits it offers. When things are difficult, having the knowledge that there will always be food available can help reduce feelings of tension and anxiety, thereby creating a sense of stability and normalcy. Preservation of food can also function as a constructive activity that creates a proactive mindset, encouraging individuals to prepare for the future rather than feeling helpless in the face of misfortune. This is because preservation of food can be a constructive activity.

A further factor that highlights the significance of food preservation in emergency preparedness is the communal character of the practice. When communities work together to preserve food, they have the potential to enhance social links and develop a sense of purpose that is shared by all members of the community. The community's food security is improved as a result of these collaborative practices, which also result in the formation of a support network that is able to effectively mobilize in preparation for and response to catastrophes. The community's collective resourcefulness is enhanced by the sharing of information and expertise concerning the preservation of food, which in turn makes the community better prepared to deal with obstacles.

On the other hand, education and training are necessary for the successful incorporation of food preservation into disaster preparedness. Individuals and communities can be equipped with the information and skills necessary to preserve food in a manner that is both safe and effective through the implementation of awareness campaigns and workshops. The education that is being provided ought to cover a variety of methods of preservation, with an emphasis on the significance of having a variety of approaches to food storage. Additionally, it should address the nutritional aspects of food preservation, making certain that the food supplies that have been preserved are able to fulfill the dietary requirements of the people during times of emergency.

In conclusion, the role of food preservation in emergency preparedness is an important one that involves a combination of several aspects. A practical answer for maintaining food security, support for independence and self-sufficiency, promotion of effective resource use, psychological comfort, and the development of community resilience are all provided by it. It is becoming increasingly clear that the preservation of food is an essential component of emergency preparedness as the globe continues to face an increasing number of uncertainties. By accepting and improving food preservation practices, society may establish a stronger foundation for dealing with calamities. This will ensure that the fundamental human need for sustenance is met with preparedness and resilience when confronted with hardship. The practice of food preservation has the potential to become an essential component of emergency preparedness methods, thereby serving as a guiding light of optimism and stability during times of crisis. This can be accomplished through public education and collaborative efforts.

Strategies for Stockpiling Pickled and Fermented Foods

When it comes to being prepared and being able to provide for oneself, stockpiling foods that have been pickled and fermented is a smart strategy that may be taken to assure both food security and nutrition. These preservation techniques, which are highly regarded for their durability, nutritional advantages, and flavor enhancement, are essential to developing a sustainable lifestyle and being prepared for unexpected events. The tactics for hoarding pickled and fermented foods are discussed in depth in this section, with an emphasis placed on the significance of these strategies in the process of constructing a resilient food supply that is capable of withstanding the unpredictability of the future.

With the help of pickling and fermentation, humanity have been able to preserve the bounty of harvests and offer food during times of scarcity. These processes have been around for centuries and have withstood the test of time. It is well known that pickled foods, which are preserved in an acidic solution, and fermented foods, which are changed by helpful bacteria, both have the ability to extend their shelf life and possess qualities that are beneficial to one's health. As a result of these traits, they are excellent candidates for storing since they provide a dependable supply of nourishment and culinary variety that can be accessed throughout the year, particularly in times of emergency.

After gaining an awareness of the fundamentals of safe preservation, the first step in hoarding pickled and fermented foods is to accumulate them. It is of the utmost importance to adhere to the specified criteria in order to guarantee that the foods are retained in the appropriate manner, hence reducing the likelihood of food deterioration and foodborne illness. This comprises the utilization of jars and equipment that have been thoroughly cleaned and disinfected, products of superior quality, and accurate proportions of salt, sugar, and vinegar. It is crucial to have a thorough understanding of the specific needs for the many varieties of pickles and ferments, as factors such as temperature and pH level can have a considerable impact on the overall quality and security of the finished product.

As soon as the basis of secure preservation techniques has been created, the next strategic focus will be on diversifying the stockpile. It is possible to pickle or ferment a wide array of foods, which includes fruits, vegetables, dairy products, and even meats, which results in a vast diversity of flavors and nutrients. Not only does diversification increase the number of culinary alternatives that are accessible, but it also guarantees that a balanced intake of vitamins and minerals is achieved. It is possible to develop a comprehensive and nutrient-dense stockpile by include a variety of pickled foods such as cucumbers, carrots, beets, onions, sauerkraut, kimchi, pickled dairy products such as yogurt and kefir, and fermented soy products which includes miso and tempeh from the fermentation process.

To ensure that pickled and fermented foods retain their quality over time, it is essential to store them in the appropriate manner. It is necessary to store the majority of pickled products in a cold, dark location in order to maintain their flavor and prevent them from going bad. Foods that have undergone fermentation, on the other hand, may call for certain storage conditions, depending on the stage of fermentation they are in. For instance, in order for certain ferments to develop their tastes, they would need to be stored at room temperature for a period of time. After that, they could need to be moved to a cooler environment in order to slow down the fermentation process. To ensure the stockpile's longevity and safety, it is essential to have a solid understanding of these conditions and to effectively manage them.

To optimize the value of the stockpile while minimizing waste, it is vital to use measures such as regular rotation and management of the stockpile. In order to accomplish this, it is necessary to maintain a record of the preserved items, including the dates on which they were prepared and the anticipated shelf life. The consumption of pickled and fermented foods on a regular basis not only contributes to the enhancement of one's diet but also guarantees that the oldest products are consumed first, so preserving a stockpile that is both contemporary and viable. When it comes to efficient stockpiling, the notion of "first in, first out" is widely recognized as a crucial principle.

Increasing the resilience and diversity of food preservation strategies can be accomplished through community engagement and the exchange of knowledge, in addition to the efforts of individuals to stockpile information. It is possible to have access to a greater variety of pickled and fermented foods by taking part in community gardens, workshops on food preservation, and local swap meetings. By exchanging information and resources, community members are able to build their relationships with one another and develop a collective sense of preparedness and food security.

In addition to ensuring one's survival in times of crisis, the advantages of storing pickled and fermented foods in reserves go much beyond that. The presence of probiotics in these meals helps to improve digestion and may also strengthen the immune system, which in turn contributes to the intestinal health of the individual. Furthermore, the rich flavors and textures that pickled and fermented foods bring to the table can provide comfort and a sense of normalcy during times of stress, underscoring the significance of taste and enjoyment in the process of nourishing oneself.

The concept of hoarding pickled and fermented foods offers a proactive approach to preparedness, specifically in light of the fact that climate change and global events are increasingly posing a danger to the stability of food supplies. This approach is consistent with sustainable living principles, meaning that it lessens reliance on outside sources of food and reduces the amount of food that is wasted. Individuals can nurture resilience in the face of uncertainty by securing a degree of autonomy over their food supply through the practice of pickling and fermenting, which requires an investment of both time and resources.

In conclusion, the act of stockpiling pickled and fermented foods is a multi-pronged strategy that involves a variety of aspects, including safety, diversification, appropriate storage, rotation, and community engagement. These foods that have been preserved are a demonstration of human creativity, since they provide a solution that is both sustainable and nourishing to the problems that arise with preserving food. By embracing these time-honored methods, people and communities have the ability to construct a robust and tasty food stockpile that is prepared to support them through whatever hardships the future may bring. This strategy not only ensures survival but also assures Not only does it improve the quality of life, but it also demonstrates that even in the face of adversity, food can continue to be a source of health, comfort, and joy.

Community Sharing and Bartering

Community sharing and trading have developed as innovative approaches to the difficulties of food security, sustainability, and social cohesion in the contemporary environment of global food systems. These reactions have emerged as a result of the evolving nature of global food systems. These methods, which have their origins in the age-old traditions of communal exchange and collaboration, provide a compelling alternative to the conventional approach to food production and distribution that is driven by the market. The practice of community sharing and bartering in the food supply not only addresses concerns of accessibility and affordability, but it also helps to develop communities that are more robust and resilient. This is accomplished by giving priority to local resources, mutual assistance, and direct exchanges. An examination of the methods, advantages, and possibilities of these behaviors is presented in this section, with a focus on the significance of these practices in the contemporary setting.

The production, distribution, as well as consumption of food within a community are all practices that are included in the concept of community sharing in the context of food supply. Community gardens, food cooperatives, and shared meals are all examples of the many different ways that this can be accomplished. These types of initiatives make it possible for members to combine their resources, divide the effort involved in the production of food, and distribute the harvest according to the level of contribution or necessity. This strategy creates a stronger relationship to the source of one's food, which in turn fosters an appreciation for sustainable agriculture techniques. It also promotes availability to fresh, healthy food, particularly in regions that are underserved by traditional retail shops.

On the other side, bartering is the practice of exchanging commodities or services directly with one another without the addition of monetary value. When discussing the provision of food, this might refer to the practice of exchanging surplus product from a home garden for other commodities or services within the community, such as the provision of home repairs, meal preparation, or child care. Bartering makes it possible to make effective use of local resources, which in turn helps to cut down on waste and ensures that any surplus food is put to good use. It also makes it possible for individuals and families to have access to a greater variety of foods and services than they might be able to buy in a conventional market environment.

When it comes to the provision of food, the advantages of communal sharing and trading are numerous. Through the provision of low-cost access to food and the reduction of reliance on cash for basic requirements, these approaches have the potential to reduce food insecurity from an economic standpoint. In terms of the environment, they promote environmentally responsible agriculture and lessen the carbon footprint that is linked with the transportation of food across great distances. By promoting a sense of mutual responsibility and support, they build the links that bind the society together on a social level. As a result of these activities, places are created for social contact, the exchange of knowledge, and the nurturing of shared ideals about environmental sustainability, health, and food.

Bartering and community sharing have hurdles in terms of scalability and legal legitimacy, despite the fact that they have these benefits. In many instances, these kinds of initiatives perform their operations on the periphery of formal economies, traversing legal frameworks that do not always make room for the exchange of food that is not for commercial purposes. In addition, although these practices have the potential to have a substantial influence on local communities, in order to scale them up to address bigger challenges with the food system, further strategic planning, resource allocation, and policy support are required.

There are a few different approaches that can be taken in order to overcome these obstacles and make the most of the opportunities that community sharing and bartering present in providing food supplies. A greater number of people can be encouraged to participate in these practices by first raising awareness and educating people about the advantages and practicalities of these practices. Providing individuals with the knowledge and skills essential to engage in food sharing and bartering can be accomplished through the organizing of community events, workshops, and demonstrations that exhibit successful models.

Secondly, the establishment of networks and partnerships among community organizations, local businesses, and government agencies has the potential to expand the resources available to food sharing and bartering projects as well as their overall reach. The establishment and expansion of these projects can be made simpler through the use of collaborations, which can make it simpler to gain access to land, finance, and logistical support.

When it comes to the third point, campaigning for policy reforms that acknowledge and support non-commercial food exchange can be of assistance in incorporating these practices into the larger food system. One example of this would be the modification of zoning regulations to make room for community gardens, the establishment of legal frameworks for food sharing, and the provision of tax incentives to enterprises that take part in bartering systems.

In conclusion, the practice of community sharing and trading in the food supply is a powerful instrument that may be utilized to improve food security, sustainability, and social cohesion. By drawing on the concepts of collaboration, mutual aid, and the utilization of local resources, these activities have the potential to assist in the development of food systems that are more resilient and egalitarian. The potential benefits of community sharing and bartering are enormous, despite the fact that there are still hurdles to be faced in terms of scalability and regulatory recognition. To pave the way for a food system that is not only more sustainable and equitable, but also deeply based in the values of community and collaboration, communities are continuing to innovate and adapt these age-old methods to modern situations. This is paving the way for a food system that is more sustainable and equitable. Through this action, they bring to our attention the fact that the concept of feeding encompasses not only the act of eating itself, but also the social connections, cultural practices, and economic transactions that are necessary for the maintenance of human communities.

CHAPTER X

Troubleshooting and FAQs

Common Issues and Solutions in Pickling and Fermentation

Pickling and fermentation are two procedures that have been an essential part of human culinary practices for thousands of years. These processes serve as a technique to preserve food, increase flavor, and boost nutritional content. However, just like any other method, they come with their own unique set of problems that might discourage practitioners of all skill levels, including those who are just starting out. Having a solid understanding of these issues and being aware of how to solve them is absolutely necessary in order to achieve good results. This section examines the typical problems that arise during the process of pickling and fermenting, and it provides remedies that can assist in overcoming these difficulties. By doing so, the author ensures that the process of preserving food continues to be both pleasurable and fulfilling.

In the process of pickling, one of the most common problems that might arise is the formation of a murky brine. This cloudiness can be brought on by minerals in the water, the deterioration of vegetables, or the growth of bacteria that are not desirable. In order to avoid this, it is possible to eliminate minerals that contribute to cloudiness by using water that has been distilled or filtered. In addition, making sure that the veggies are clean and sliced in a regular manner might help reduce the amount of sediments that are released into the brine and its subsequent breakdown. In the event that cloudiness continues, it may be an indication of the growth of microorganisms. In this scenario, ensuring that the appropriate proportions of vinegar, water, and salt are present and adhering to sterile practices will help prevent contamination.

One of the difficulties associated with pickling is that it causes the veggies to become softer, which prevents them from achieving the proper crispness. This problem is frequently the result of enzymatic processes that take place within the plants. Either adding grape leaves, which contain tannins that help preserve crunchiness, or quickly soaking veggies in ice water before pickling them is a frequent remedy. Grape leaves also contain tannins. In addition, it is essential to make certain that the pickling process starts with veggies that are fresh and hard, as older fruit has a greater tendency to become softer.

When it comes to fermentation, one of the most significant issues is the development of mold or yeast on the surface, which can take place as a result of the product being exposed to air. Aside from the fact that it is unpleasant, this matter may also be hazardous to one's health. The answer is in the creation of an anaerobic environment, which is an atmosphere devoid of oxygen. This can be accomplished by the utilization of airlock systems or by making certain that the vegetables that are fermenting are entirely submerged in brine. In order to prevent the exposure that might contribute to the growth of mold on vegetables, it is possible to use fermenting stones or weights to keep the veggies concealed beneath the liquid.

During the fermentation process, one of the most prevalent problems that might arise is the formation of an unpleasant stench, which can deter many people from continuing their fermentation activities. It is typical for there to be some odor, particularly in the first stages of fermentation; however, aromas that are overly nasty can be an indication of spoiling or the presence of bacteria that are hazardous. It is possible to alleviate this issue by ensuring that the correct salt content is maintained, which prevents the growth of undesirable microorganisms, and by keeping the fermentation temperature at an appropriate level. In the event that an unpleasant odor continues to be present, it is recommended that the batch be thrown away and that a new batch be made with a strict attention to cleanliness and proportions.

Another difficulty that might arise in the process of pickling and fermentation is the fact that there is a wide range of flavor and texture, which can result in inconsistent outcomes. Some of the reasons that contribute to this variability include variances in the quality of the ingredients, environmental circumstances, and the precision of the measurements. When it comes to achieving consistent results, it is absolutely necessary to make use of the freshest possible components and to adhere to recipes very precisely, particularly with regard to the proportions of salt, sugar, and vinegar. By controlling the environment in which fermentation takes place, including the temperature and humidity, it is possible to help standardize the process, which will ultimately lead to results that are more predictable and satisfying.

Along with these particular challenges, one of the more general difficulties associated with pickling and fermentation is the possibility of contamination by bacteria that are toxic to the food. It is of the utmost importance to maintain hygiene during the process. This includes properly cleansing hands and surfaces, sterilizing jars and equipment before use, and using water and ingredients of a high quality that are free of contamination. By taking these measures, the risk of contamination can be considerably reduced, which will ensure that the pickling and fermentation processes are carried out in a safe and successful manner.

The benefits of pickling and fermenting are vast, delivering a wide variety of aromas, textures, and nutritional advantages. On the other hand, these processes are not without their difficulties. Perseverance as well as a willingness to learn from one's mistakes are essential qualities for individuals who are experiencing hardships. Participating in a network of other enthusiasts, whether through online forums, workshops, or local clubs, can provide further support, inspiration, and advice on how to solve problems.

In conclusion, although the typical problems that arise throughout the fermentation and pickling processes can be disheartening, they are not insurmountable. It is possible for anyone to become proficient in these time-honored methods if they have a fundamental comprehension of the factors that are responsible for these issues and if they put into practice remedies that are applicable. Paying close attention to the details, keeping a clean environment, and being willing to try new things are the most important things to do. Through this method, the skill of pickling and fermenting can continue to be a gratifying aspect of the culinary practice, allowing us to connect with our history while also providing us with nourishment in the present.

Answers to Frequently Asked Questions

Pickling and fermentation are both practises that have been around for centuries, but in recent years, there has been a resurgence of interest in them due to the fact that they are beneficial to one's health, they are environmentally friendly, and they lend distinctive flavours to food. An increasing number of people are beginning the path of making pickled and fermented foods at home, which has resulted in the emergence of a multitude of frequently asked questions that represent the common concerns and curiosity that people have regarding these procedures. It is the purpose of this section to respond to these questions and provide answers that are thorough in order to assist both novices and seasoned aficionados in their explorations of the culinary world.

The inquiry "What is the difference between pickling and fermentation?" is one of the most often asked questions. Both of these techniques are considered to be ways of food preservation; yet, they function through different mechanisms. Pickling is the process of preserving food by drowning it in an acidic liquid, most often vinegar, in order to provide an environment that inhibits the growth of bacteria. Fermentation, on the other hand, is a process that relies on the action of helpful bacteria to convert sugars in food into alcohol or acids. This process naturally preserves the food while also improving its flavors and nutritional value. It is essential for everyone who is interested in delving into these practices to have a solid understanding of this fundamental distinction, since it has an impact on the components, procedures, and conditions that are necessary for successful preservation.

Another question that is asked rather frequently is, "What can I do to make sure that my pickles maintain their crispness?" This is a common source of dissatisfaction for people who are new to pickling since they wind up with pickles that are soft and soggy rather than the crunchy texture that they desire. The key to making crisp pickles resides in a number of aspects, including the freshness of the product, the use of salt that has not been iodized, and the addition of substances that enhance crispness, such as grape leaves or calcium chloride. It is possible to considerably influence the final texture of the veggies by ensuring that they are freshly collected and prepared as rapidly as possible.

Concerns about safety frequently prompt people to ask, "How can I tell if the fermented foods I have in my possession have gone bad?" It is vital to have both visual and smell signals in order to determine whether or not a fermentation process was successful. The presence of a sour odor, a modest effervescence, and the absence of mold or other unpleasant odors are all indicators that the fermentation process was not unsuccessful. Mold, a scent that is obviously unpleasant, or a texture that is slimy are all indications that the batch must be rejected since it has ruined and should be thrown away. When it comes to determining whether or not fermented foods are safe to consume, it is essential for individuals to rely on their own senses and err on the side of caution.

One of the most popular questions that fans have is, "Is it possible to ferment any kind of vegetable?" It is possible to ferment the majority of vegetables; however, certain vegetables are more suited to the procedure than others. Cucumbers and cabbage, which have a greater water content, are classic favorites because they ferment well and create rich flavors. Other vegetables that have a higher water content include kale and spinach. Experimenting with different kinds of vegetables, on the other hand, can produce delicious outcomes, which can encourage people to go beyond the options associated with standard cuisine. To get the desired results, it is essential to modify the salt concentrations and fermentation times in accordance with the particular features of each vegetable.

Concerns such as "Why does my fermented food taste too salty?" are frequently voiced by individuals. When it comes to fermentation, maintaining a healthy balance of salt is essential since it prevents the growth of bad bacteria while allowing helpful bacteria to flourish. If a batch has an unusually salty flavor, it may be because the amount of salt that was measured was excessively high or because the fermentation time was not long enough.

It is possible that the saltiness that is perceived will decrease as fermentation occurs because the lactic acid bacteria release acids that bring the flavors into balance. In order to overcome this problem, you can either let the fermentation process continue for a longer period of time or make adjustments to the salt ratio in subsequent batches.

The question "Do I need special equipment to start pickling or fermenting?" is another one that comes up after the first one. The simplicity of pickling and fermenting, as well as the fact that they require very little equipment, is what makes them so appealing. It is sufficient to begin with the most fundamental kitchen utensils, which include jars, lids, bowls, and measuring spoons at the very least. When it comes to fermentation, the utilization of weights to immerse the veggies in the brine and the utilization of airlock lids are both things that can help establish an anaerobic atmosphere; nevertheless, these are not strictly necessary for novices. It is possible that individuals will decide to purchase specialist equipment in order to improve their experience as they become more active in these practices.

This brings us to our final question, which is frequently asked: "How long can I store fermented and pickled foods?" Depending on the method that was used, the ingredients that were used, and the conditions under which the item was stored, the shelf life of fermented and pickled foods might vary. Generally speaking, pickles that have been properly sealed can be kept for up to a year if they are kept in a cold and dark location. Fermented foods that are kept in the refrigerator after fermentation can be preserved for a number of months, and the flavor of certain types of fermented foods can even improve over time. Jars should be dated and labeled, and it is essential to check stored goods on a frequent basis for any indications that they may have gone bad.

In conclusion, the technique of pickling and fermentation opens up a world of culinary possibilities thanks to the exquisite flavors, health advantages, and deeper connection to ancient ways of food preservation that they provide. Individuals are able to embark on this journey with confidence knowing that they are equipped with the knowledge necessary to handle the complexities of these practices if they address the typical questions and concerns that they have. Patience, experimentation, and a desire to learn from each batch are essential components in the process of learning the art of pickling and fermentation, just as they are in any other culinary endeavor.

CONCLUSION

Recap of Key Points

Throughout the pages of "Pickling and Fermentation for the Prepared: Creating a Resilient Food Supply," we have explored the rich tapestry of pickling and fermentation, uncovering a wealth of knowledge, techniques, and insights that empower us to navigate an uncertain world with confidence and creativity. As we recap the key points of our journey, it becomes clear that pickling and fermentation are not just culinary techniques; they are powerful tools for building resilience, promoting sustainability, and fostering community connections.

First and foremost, we have learned that pickling and fermentation are ancient food preservation methods that have stood the test of time. From the pickled cucumbers of ancient Mesopotamia to the sauerkraut of medieval Europe, humans have been harnessing the power of these techniques for centuries to expand the shelf life of perishable foods and enhance their flavor and nutritional value. Understanding the historical context as well as cultural significance of pickling and fermentation gives us a deeper appreciation for their enduring relevance in today's world.

At the heart of pickling and fermentation lies a fundamental understanding of the science behind these processes. Whether through the creation of acidic brines in pickling or the metabolic activity of microorganisms in fermentation, the transformation of raw ingredients into flavorful, shelf-stable delicacies is driven by natural chemical reactions. By grasping the basic principles of pH, microbial activity, and enzymatic processes, we can troubleshoot common issues, experiment with new techniques, and unlock the full potential of pickling and fermentation in our own kitchens.

In addition to their practical applications, pickling and fermentation offer myriad benefits for individuals and communities. From reducing food waste and promoting sustainability to enhancing food's nutritional value and digestibility, these techniques represent holistic approaches to food preservation and consumption. By harnessing the power of pickling and fermentation, we can create nutrient-dense foods that support digestive health, strengthen immune function, and promote overall well-being.

Furthermore, pickling and fermentation provide opportunities for culinary creativity and exploration. From classic recipes like pickles and sauerkraut to exotic creations like kimchi and kombucha, the world of pickled and fermented foods is as diverse as it is delicious. By experimenting with different flavors, textures, and techniques, we can expand our culinary horizons, discover new favorites, and share our creations with friends and family.

But perhaps most importantly, pickling and fermentation are about more than just food; they are about community resilience and connection. Whether through sharing surplus harvests with neighbors, bartering homemade preserves with friends, or preserving family recipes for future generations, pickling and fermentation bring people together around the shared joys of food and fellowship. In an increasingly fragmented and uncertain world, these age-old practices remind us of the importance of community, cooperation, and mutual support.

In conclusion, "Pickling and Fermentation for the Prepared" is not just a book about pickles and kraut; it is a roadmap for building resilience, promoting sustainability, and fostering community connections in an uncertain world. By embracing the lessons learned and the experiences shared in these pages, we can create a more sustainable, secure, and satisfying future—one jar of pickles at a time.

Encouragement for Further Exploration

As we conclude our journey through the world of pickling and fermentation, it's essential to recognize that our exploration is far from over. In fact, it's just the beginning. The techniques and insights we've gained in "Pickling and Fermentation for the Prepared: Creating a Resilient Food Supply" serve as a foundation for further discovery and experimentation, inviting us to delve deeper into the rich tapestry of flavors, traditions, as well as possibilities that await.

One of the most exciting aspects of pickling and fermentation is the endless potential for creativity and innovation. From exploring new flavor combinations to experimenting with different fermentation vessels and techniques, there are countless opportunities to put our newfound knowledge into practice and create unique culinary masterpieces. Whether you're a seasoned home cook or a curious novice, the world of pickled and fermented foods offers something for everyone to explore and enjoy.

Moreover, the journey of pickling and fermentation is not just about what happens in the kitchen; it's about connecting with the wider world around us. By seeking out local ingredients, supporting small-scale farmers and producers, and engaging with community-based initiatives, we can deepen our understanding of the food we eat and the systems that sustain us. From attending farmers' markets and food swaps to participating in workshops and classes, there are countless ways to engage with our local food ecosystems and cultivate meaningful relations with the people and places that nourish us.

In addition to fostering connections with our local communities, pickling and fermentation also offer opportunities for global exploration and cultural exchange. From sampling traditional pickled delicacies from around the world to learning about their history and traditions, there is much to discover and appreciate in the diverse array of pickled and fermented foods that grace our tables. By embracing a spirit of curiosity and openness, we can expand our culinary horizons and develop a more profound appreciation for the richness and diversity of global food cultures.

But perhaps most importantly, the journey of pickling and fermentation is about personal growth and self-discovery. As we experiment with new techniques, overcome challenges, and celebrate successes, we develop confidence in our abilities and a sense of pride in our accomplishments. Whether mastering the art of lacto-fermented pickles or perfecting the balance of flavors in a batch of homemade kimchi, each step brings us closer to realizing our full potential as cooks and creators.

In conclusion, "Pickling and Fermentation for the Prepared" is not just a book; it's a call to action to explore, experiment, and embrace the transformative power of pickling and fermentation in our lives. By continuing our journey of exploration and discovery, we can unlock new flavors, forge new connections, and create a more resilient, sustainable, and satisfying future for ourselves and our communities. So, let us venture forth with curiosity and courage, knowing that the possibilities are endless and the rewards are boundless.

Final Thoughts on Creating a Resilient Food Supply

As we reflect on our journey through the world of pickling and fermentation, it's clear that the principles and practices we've explored extend far beyond the kitchen. At its core, the quest for a resilient food supply is about more than just preserving fruits and vegetables; it's about reclaiming control over our food systems, reducing waste, and fostering a deeper connection to the foods we consume. By embracing the time-honored techniques of pickling and fermentation, we empower ourselves to create a more sustainable, secure, and satisfying future for ourselves and future generations.

One of the key lessons we've learned is the importance of self-reliance and preparedness in an uncertain world. By learning how to preserve and store our own food, we reduce our reliance on external sources and gain greater control over our food supply. Whether it's stocking our pantries with homemade pickles and ferments or sharing surplus harvests with neighbors, each step we take towards self-sufficiency brings us closer to establishing a more resilient food system that can withstand the difficulties of an ever-changing world.

Moreover, the journey of pickling and fermentation is about more than just preserving food; it's about preserving tradition, culture, and heritage. From the sauerkraut of Eastern Europe to the kimchi of Korea, pickled and fermented foods are integral components of diverse culinary traditions around the world. By learning about the history and cultural significance of these foods, we gain a deeper appreciation for the wisdom and ingenuity of generations past who understood the value of preserving seasonal abundance for leaner times.

In addition to preserving tradition, pickling and fermentation also offer opportunities for innovation and creativity. By experimenting with different flavors, techniques, and ingredients, we can put our own unique stamp on age-old recipes and create culinary masterpieces that reflect our individual tastes and preferences. Whether it's infusing pickles with exotic spices or fermenting unusual vegetables, the possibilities are limited only by our imagination.

But perhaps most importantly, the journey of pickling and fermentation is about building community resilience and fostering connections with the people and places that sustain us. By sharing our knowledge, skills, and surplus harvests with others, we create networks of mutual support and cooperation that strengthen our communities and enhance our collective well-being. Whether it's participating in food swaps, joining community gardens, or volunteering at local food banks, each act of solidarity brings us closer together and builds a more resilient, compassionate society.

In conclusion, the journey of pickling and fermentation is a journey of empowerment, connection, and transformation. By embracing the principles as well as practices outlined in "Pickling and Fermentation for the Prepared," we can create a more sustainable, safe, and satisfying future for ourselves and future generations. So let us continue to explore, experiment, and innovate, knowing that the rewards of resilience and self-reliance are well worth the effort.

Thank you for buying and reading/ listening to our book. If you found this book useful/ helpful please take a few minutes and leave a review on the platform where you purchased our book. Your feedback matters greatly to us.

www.ingramcontent.com/pod-product-compliance
Lightning Source LLC
Chambersburg PA
CBHW072013150726
47999CB00002B/639